Computer-Aided Design in Magnetics

D. A. Lowther   P. P. Silvester

Computer-Aided Design in Magnetics

With 84 illustrations

Springer-Verlag
Berlin Heidelberg New York Tokyo

D. A. Lowther
Associate Professor
 of Electrical Engineering
McGill University
Montreal H3A 2A7
Canada

P. P. Silvester
Professor of Electrical
 Engineering
McGill University
Montreal H3A 2A7
Canada

Library of Congress Cataloging in Publication Data
Lowther, D. A.
  Computer-aided design in magnetics.
  Bibliography: p.
  Includes index.
  1. Magnetic devices—Design and construction—Data
processing. 2. Computer-aided design. I. Silvester,
P. P. II. Title.
TK454.4.M3S55   1985   621.34   85-17227

Media Conversion and Typesetting by House of Equations, Newton, New Jersey.
Printed and bound by Halliday Lithograph, West Hanover, Massachusetts.
Printed in the United States of America.

9 8 7 6 5 4 3 2 1

ISBN 3-540-15756-5 Springer-Verlag Berlin Heidelberg New York Tokyo
ISBN 0-387-15756-5 Springer-Verlag New York Berlin Heidelberg Tokyo

Preface

Computer-aided design has come of age in the magnetic devices industry. From its early beginnings in the 1960s, when the precision needs of the experimental physics community first created a need for computational aids to magnet design, CAD software has grown to occupy an important spot in the industrial designer's tool kit.

Numerous commercial CAD systems are now available for magnetics work, and many more software packages are used in-house by large industrial firms. While their capabilities vary, all these software systems share a very substantial common core of both methodology and objectives. The present need, particularly in medium-sized and nonspecialist firms, is for an understanding of how to make effective use of these new and immensely powerful tools: what approximations are inherent in the methods, what quantities can be calculated, and how to relate the computed results to the needs of the designer. These new analysis techniques profoundly affect the designer's approach to problems, since the analytic tools available exert a strong influence on the conceptual models people build, and these in turn dictate the manner in which they formulate problems. The impact of CAD is just beginning to be felt industrially, and the authors believe this is an early, but not too early, time to collect together some of the experience which has now accumulated among industrial and research users of magnetics analysis systems.

In its early versions, this book has been used by students in the authors' one-semester course at McGill University. It has also served as the cornerstone for numerous short courses given to industrial audiences. The authors wish to acknowledge their indebtedness to Dr. G. K. Cambrell, for his meticulous reading of the manuscript, and to Dr. E. M. Freeman, for our many discussions and his contributions to an early draft of several chapters. Particular thanks are due, however, to the many colleagues and students whose suggestions, help, and numerous examples have shaped this book.

D. A. Lowther
P. P. Silvester

Contents

Field Problems in Magnetic Devices   119

Postprocessing Operations in CAD   160

Postprocessing Systems for Magnetics   228

CAD Systems Present and Future   258

The Literature of CAD in Magnetics   293

Index   321

Introduction

Magnetic devices have traditionally been designed by combining empirical rules based on experimental evidence with simplified magnetic circuit models. This technique may be labelled *design by rule*. But as devices become increasingly varied and complex, conventional design rules are no longer adequate and *design by analysis*, based on reasonably detailed solution of the underlying electromagnetic field problems, becomes the normal practice. Design by analysis really means design by numerical analysis since no other tools are capable of dealing with both the geometric complexity and nonlinearity found in such disparate devices as vertical recording heads and direct current machines. Design by analysis therefore inevitably means at least computer-aided analysis, and increasingly has come to mean full-fledged computer-aided design.

This book outlines what can and what cannot be done with the computer-aided analysis systems now available commercially. Its main purpose is to show how the new techniques can be brought to bear on design problems; how design rules or novel approximations can be developed for new situations; and how CAD system capabilities can be stretched to cover problems that the software designers may not have had in mind. It assumes that analysis methods for electromagnetics problems are now sufficiently well known and addresses itself largely to the engineering task of formulating electromagnetics problems in a computable fashion. It is principally, indeed almost exclusively, concerned with two-dimensional analysis, for two good reasons. First, nearly all commercially available CAD software deals with two-dimensional problems. Secondly, and more importantly, learning to use three-dimensional analysis tools is a very great deal more difficult that working in two dimensions; it pays, as the saying goes, to learn to walk before you run.

Analysis Methods in Electromagnetics

The techniques used for electromagnetic device analysis in practical design may be divided into two classes: the traditional techniques, which rely on notions of magnetic circuit analysis, and the newer numerically implemented approaches. Both have been used in computer-aided design systems, but there is little doubt that the modern field theoretic techniques are gaining the upper hand. This book is concerned primarily with such new techniques, and with ways of marrying the old with the new.

Traditional Design Methods

Traditional development, analysis, and design of electromagnetic devices have tended to use extremely simple analytic models, supplemented by experimental evidence. Most of these consist of electric or magnetic equivalent circuits postulated on the basis of experience and intuition. Frequently, such simple approaches lead to qualitatively correct but quantitatively inaccurate results, which are supplemented with correction parameters, thus leading to *design rules*. The simplified models and rules so developed have long provided, and continue to provide, designers with data concerning specific device features such as the input impedance, transient and subtransient reactances, power output and efficiency. However, they are usually restricted to very narrow classes of devices, because the various numerical correction factors are empirically introduced, and can be relied on only for designs substantially similar to those which furnished the experimental evidence. Occasionally, more sophisticated analytic techniques derived from basic electromagnetic principles have been used when a more detailed insight has been required. But these are labor intensive and require high levels of expertise in electromagnetics and mathematics. The traditional methods have the advantage, from the design engineer's viewpoint, of being relatively easy, and fast to apply. In quite a few cases, they can be expressed in the convenient form of design tables or curves.

The principal shortcoming of traditional design methods lies in their inability to accommodate new situations. New magnetic materials, both hard and soft, can place additional restrictions and at the same time open new possibilities. Legal requirements relating to electromagnetic interference, safety, noise, and pollution levels lead to new forms of design. Other considerations, such as weight and size, may need detailed knowledge of the field distribution within the device. And last but not least, there are many devices which have never been built before: perpendicular recording heads and magnetic print heads are just two particular examples!

Numerical Methods

The advent of the digital computer, and its subsequent development, provided engineers with a tool capable of manipulating large volumes of data at very high speeds. In the three decades since computers became common, methods have been developed for solving the electromagnetic field equations for a great range of geometries and materials. The finite element technique in particular has proved particularly flexible, reliable, and effective in the hands of nonspecialist users.

For the design engineer these new computational tools could provide solutions to problems with a greater degree of accuracy than was previously possible. During the decade of the 1970s, conventional wisdom held that the availability of good computational methods had turned the clock back a hundred years, to that golden era in which magnetic devices were designed almost exclusively by solving the underlying magnetic field equations—until the range of analytically solvable problems was exhausted early in the twentieth century. In fact, history may well be repeating itself a century later. Two-dimensional (and some three-dimensional) problems are now relatively easy to solve yet still difficult enough to preclude the exclusive use of finite element programs in preference to classical rule-based design. Indeed, a major use of the new CAD tools will very likely be in deriving new, more sophisticated design rules. The new element for the designer may well become the *disposable design rule*, created for a relatively narrow class of designs as the result of extensive computer simulation. The narrow range of applicability of a highly specialized design rule is not a problem in an era of cheap computing, in which new rules can be created by additional quick simulations as needed.

The System Life-Cycle

All engineering software is necessarily highly specialized. Its content is shaped by the needs of potential users as the system designer perceives them, by the analytic techniques available when the system is designed, and by the computer hardware in reasonably common use at the time. All three factors change with time at a greater or lesser speed, so that the life-span of any software package is limited.

Good engineering analysis software takes time to design and perfect. A typical large-scale finite element package involves tens of thousands of lines of code, taking man-years to design and test. Such an analysis package normally requires 2–3 years to create and must remain on the market for 7–10 years if its high initial cost is to be justified. Its total life-cycle is therefore about 10–12 years, a long time in a high-technology world.

Hardware Development

Currently, major changes in computer hardware occur about every 3–4 years. This rate of change applies equally to processor hardware and to input-output devices. Changes in processor hardware are often unspectacular from an analyst's point of view, since they usually result in established operations being carried out more conveniently, quickly, or cheaply. Changes in input-output devices, on the other hand, tend to have a more revolutionary aspect, since they frequently alter the external appearance of systems fundamentally. For example, the introduction of graphics terminals in the 1970s amounted to more than just an increase in plotting speed: it altered the way work was done.

Since the economic life of software systems is substantially longer than the hardware life-cycle, every analysis system must reside on obsolete hardware at least half its lifetime. Conversion to different hardware systems is commonly done, but it is at best a half-measure, for the software structure is influenced very strongly by what the hardware is capable of doing. For example, a light pen and a graphics tablet have a great deal of superficial similarity in use, but the tablet can be used to digitize drawings and curves directly, while the light pen cannot. Software transported from light pen environments to graphics tablets therefore ends up merely using the tablet to mimic a light pen, without taking advantage of its additional abilities. As another very familiar example, keyboard terminals fully capable of writing both upper and lower case characters were for years (and in many systems still are!) used to display capital letters only, because the software systems driving them were originally written for terminals incapable of printing lower case characters.

Transportation of software systems to new hardware usually means making the new hardware act like the old. Such mimicry is often easy, in the sense that the newer devices normally provide all the capabilities of the old and furnish new functions besides. But alteration of software to make good use of new hardware capabilities may imply not just a few alterations but basic redesign. Ignoring the capabilities of new hardware is thus possible for a few years, but only for a few, because the unused hardware facilities eventually come to predominate over those actually exploited by the programs. A new start is then indicated.

A very rapidly evolving area, in which most current packages leave much to be desired, is that of man–machine communication. To be fair and charitable to package designers, this area is very strongly dependent on terminal hardware, a particularly fast-changing part of an exceptionally fast-moving industry.

The Software Environment

A problem at least as vexing as rapid hardware evolution is the continual and rapid change of system software in which any CAD system must live. Operating systems are subject to even more updates, alterations, and corrections than hardware.

Because elementary operations in arithmetic processing, all input–output operations, and many other processes are hardware-dependent, they normally call on machine-language routines resident in the operating system software. Indeed, the whole point of operating systems is to isolate the user from too intimate an involvement with the hardware details of the machine. Computer users—even experienced programmers—often do not appreciate quite how major a part of machine action is really not carried out by the programs themselves but turned over to parts of the computer operating system instead. For example, it takes a Fortran program of only six or seven lines to read a pair of input numbers, multiply them, and print the answer. When finally ready to run, such a program will have grown to dozens of kilobytes in size, for it will call on perhaps fifty or more utility programs contained in the operating system. With a different operating system, the same hardware may behave very differently; it may be more responsive in interactive work, yet slower in grinding through heavy arithmetic tasks. Because many operating systems are machine-specific, users may confuse the characteristics of the machine with the characteristics of its operating system. They often speak of the peculiarities of "talking to the VAX-11" when they really mean "talking to the Unix operating system". In fact, the VAX-11/750 machine will present a totally different appearance at the user terminal if run with the VAX/VMS operating system instead of Unix, and conversely an HP-9000 running under Unix will pretty much resemble the VAX.

Traditionally, computer makers have furnished system software in a single package with the machines themselves. In the 1960s, it was conventional to write a new operating system for every new machine. But operating systems are software systems, subject to much the same laws as CAD systems, and a life-span far exceeding hardware lifetime is necessary to make them viable. In the 1970s it became common to write operating systems for families of machines, and the 1980s are witnessing the appearance of more or less machine-independent operating systems.

Transporting a CAD system from one operating system environment to another is a difficult task. Recent CAD systems have stressed modularity and system-independence, usually at some sacrifice in memory economy and speed. Older systems, initially intended for simpler operating systems and smaller computers, were often built to obtain good performance by exploiting both hardware and operating system peculiarities. Such systems are particularly difficult to move, and particularly likely to malfunction when taken out of their native environments.

Analysis Techniques

New numerical analysis methods likely to affect engineering software have appeared at intervals of about ten or twelve years. For example, the large systems of simultaneous equations which commonly arise in continuum analysis were solved by relaxation methods in 1955–1965, by sparsity-exploiting Gaussian elimination techniques in 1965–1975, and by preconditioned conjugate gradients since about 1975. Since analytic techniques change in time-spans comparable to those of software systems, software generally lags about half a generation behind.

Alteration of existing programs to employ new analytic methods is much less easy than it looks at first glance, just as hardware changes can be fully exploited only by redesign of software. To pursue the above example, iterative or semi-iterative methods (i.e., relaxation and conjugate gradient methods) are essentially solution improvers rather than solvers: they construct a sequence of increasingly accurate solution estimates beginning from some initial estimate and can therefore exploit whatever prior knowledge may be available about the solution. In particular, they are very efficient at computing sequences of closely related solutions, using each one as the initial guess for the next. Sparse Gaussian elimination is often better at computing a single solution, but it cannot make efficient use of a priori information. Consequently, well designed software based on Gaussian elimination will be designed to let the user specify isolated problems, while systems based on conjugate gradient methods should encourage the user to generate families of successive solutions closely related to each other. Indeed, it might be argued that an ideal software system based on iterative methods should examine the problems presented by the user and sequence them so as to form a continuous curve in problem space. The value of such sequencing depends on the use of a particular kind of solver methodology, illustrating the point that the external characteristics as seen by the user will be different in systems using different methods, even though it may not be at all evident when and where these methods are used internally.

The choice of analytic technique is strongly dependent on current processor hardware. To continue the above example, a major contributing factor to the entry of Gaussian elimination techniques in the 1960s was the replacement of magnetic tape by random-access disk drives as the normal secondary storage medium. The subsequent popularity of gradient methods owes a good deal to the availability of abundant cheap memory. Indeed, one might well surmise that the 1990s will see a resurgence of interest in Gaussian elimination, since such techniques ought to reap great benefits from the development of cheap array processors in the 1980s.

The System Designer's View

A very important factor in determining the characteristics of a CAD system is the system designer's understanding of the user's requirements. It is an axiom of the software design art that the worst way to determine the user's needs is to ask him, because the user's own assessment is conditioned too strongly by tools available in the past. The user's own perception of need is usually not problem-motivated, but solution-driven, for most engineering analysts are successful solution-appliers, not problem-solvers. Hence, users of CAD systems are generally quite good at specifying what fine-tuning alterations to a system might be desirable, but not very good at defining the overall structure or functions of a system. Just the contrary is often true of system designers.

The relatively long lifetime of software systems in part derives from the large investment involved in system design and construction, but perhaps it may reflect even more the high cost of user training. The issue at hand here is not operator training, which deals with the details of system operation, but rather the acquisition of enough specialist engineering knowledge to use the results effectively. In other words, users need to develop an easy familiarity with magnetic and electromagnetic fields, as well as with practical techniques based on field theory. Good current software packages take days or weeks of operator training, but months, even years, of user training. Electromagnetic engineering relied heavily on field-theoretic concepts in the late 1800s and early 1900s, then turned to a circuit-theoretic viewpoint in the 1920s and 1930s. Half a century later, fields-based skills need to be rediscovered and applied to new situations, a difficult and time-consuming process.

Computing Equipment for CAD

The usefulness of a numerical technique to the designer depends on both the ease of implementing it on the available computer hardware and the form of access the user must have to the hardware in order to use the software effectively. Computing machinery has developed rapidly with improvements in semiconductor technology, and a large spectrum of equipment is now available from the microprocessor to the supercomputer. The selection and wise use of this immense range of resources is a prime matter of concern for both designers and users of CAD systems.

Appropriateness and Accessibility

For the user, computer systems have two characteristics of critical importance: their appropriateness to the task at hand and their accessibility when a job needs to be done. These two factors do not necessarily imply similar requirements; in fact, they may well be in direct contradiction.

Appropriateness for design usually implies a high degree of interactivity in computer use. The design of an electromagnetic device by classical methods is a true interactive process. The designer has access to all relevant design rules and procedures, can refer to these quickly, and does do so frequently as the design progresses in a back-and-forth exchange between conflicting requirements. To be of use, a computer-based design system should provide the user with either faster access to the rules and data previously used, or better data and more refined design rules, or, preferably, both. Classical batch-processing systems can permit access to rich data bases, but they cannot provide the rapidity of work which designers often desire.

Accessibility is very much a function of the available hardware and the practices of the user organization. Many small computers and design workstations are cheap enough to justify purchase for a single objective. But the organizational problems of overhead costs, floor space for yet another machine, and keeping track of yet another maintenance contract, cost effort as well as money, so there is a strong incentive to minimize the number of separate hardware systems. Sadly, there is as yet little standardization in the computing world; corporate computing in most organizations began with accounting and management, and engineering computing was for many years forced into the same batch-processing mold. Although the computer shared between management and engineering is increasingly a rarity, traditions die hard and much engineering computing remains in that management-based world of batch processing where it began. Accessibility in such cases may well conflict with appropriateness.

Ease of access, in both the hardware and organizational aspects, is a prime consideration for the design engineer. As a practical matter, tools which are easy to use are often used because they are easy to use; conversely, inaccessible tools are often hardly used at all, because inconvenience breeds unfamiliarity which in itself forms a major barrier. In order for design software to be used on a day-to-day basis, it is important to have it readily available to the engineer, thus suggesting at least partial use of low-cost, low-power, micro- or minicomputer-based systems in which maximizing the hardware utilization is secondary to effective use of human labor.

Workstations for Interactive Computing

The all but conflicting requirements of appropriateness and accessibility are difficult to meet without compromise in a single computing installation. Most users appear to value interactivity highly and seem willing to sacrifice computing speed and file access convenience if necessary. As a result, the use of interactive workstations is becoming increasingly common. These typically provide the design engineer with some local processing power and a reasonable amount of disk storage. Most importantly, however, they are equipped with graphics devices which present results in a format to which the user is accustomed. These have decreased in cost in much the same way as processors and, as a consequence, high resolution (512 by 512 dots or better) color displays are widely available. Thus, a typical workstation might consist of the following hardware:

Central processor with suitable memory
Raster display (monochrome or color)
Graphics tablet or other input device
Alphanumeric terminal
Disk storage

It would often be desirable to add

Hardcopy unit
Local area network interface

Such hardware workstations currently cost between 25% and 100% of the annual salary of a design engineer and are likely to become cheaper with time. During the 1975–1985 decade prices decreased at 20% or more annually, and it appears they will continue to do so for some time to come.

Workstations of the above hardware configuration are capable of running extensive analysis programs while at the same time providing a high degree of interactivity with the user. The local area network interface provides a high speed link (of the order of 1–10 megabits/second) to other processors. Thus, tasks can be dispatched to the processor most suited to execute them. For example, the solution of a large equation system may best be done on a large processor (e.g., a supercomputer) attached to the network, while the initial data preparation, and subsequent result interpretation, may be more appropriately handled by a smaller local processor equipped with good display hardware.

Networking

It is argued by some that the requirements of designers would be best served by computers suitable for the entire range of engineering computing tasks. Magnetic, electrical, thermal, stress, and any other necessary design or analysis packages should be usable on the same computing system, preferably the same one as is used for engineering management needs such as task scheduling and bills of materials. With standardization incomplete but progressing, there is some hope of achieving this goal through computer networks.

Many engineering computing tasks represent mixtures of high interactivity and good graphics (as in reviewing the geometric shape and layout of a proposed design) with very demanding arithmetic processing—say the solution of 100,000 simultaneous equations. In fact, engineering computing might be characterized as a rather granular mixture of these two extremes; there are not many jobs where moderate communications and graphics requirements are coupled with concurrent and equally moderate demands for central processor power. Networking promises to give the designer access to computing systems that come close to this ideal by making a *supercomputer*, a really powerful arithmetic manipulator, collaborate with a *graphic workstation*, a small computer endowed with excellent provisions for interactive work.

The modern supercomputer is an extension of the earlier computing machines in which each advance in technology has been used to increase computing power while keeping system cost approximately constant. These machines are expensive; it is therefore important to keep their utilization high. This economic constraint leads to an effort to optimize the use of internal processing facilities at the expense of input-output operations. In fact, supercomputers are designed to talk only to other computers and rarely possess any facilities for communication with a human user.

Interactive workstations are based on microprocessor technology and are in a sense opposite to supercomputers. In devising microprocessors, advances in technology have been used to reduce cost while keeping computing power roughly constant. They are cheap, so that workstation design seeks to optimize the use of human labor by stressing effective man-machine communication; high utilization of the computing hardware is not a consideration.

In the ideal future world of networking, the designer will employ an interactive workstation for all communication-intensive work, and a supercomputer for all computation-intensive work. Indeed, he can do so today; but the necessary job sharing between machines is not taken care of automatically—the designer himself must decide and instruct the computers which is to do what. Few CAD system users wish to become sufficiently expert in computer technology to make the decisions involved, to learn the command languages for controlling the network as well as the individual computers, and generally to be diverted from their own tasks. The wide application of networking in CAD therefore awaits the next generation of operating systems and the hardware they support.

Finite Element Analysis Systems

Of the numerical analysis methods available for solving the electromagnetic field equations, the finite element method undoubtedly has achieved preeminence. In fact, every general-purpose analysis package now available uses the finite element method for its mathematical operations. These are reviewed briefly in the chapter entitled *CAD Systems Present and Future*. Unfortunately, these descriptions are at best approximate, for CAD is a fast-changing area and details of any particular analysis system become outdated very quickly indeed. The vast majority of proprietary CAD systems used intramurally by major electrical firms is based on finite element methods as well, at least so far as published information is available about them. The General Electric Company and Westinghouse Electric Corporation in the United States base almost all their analysis programs on finite elements; so do Toshiba and Hitachi in Japan, Philips in the Netherlands, Brown Boveri in Switzerland, and many other leading firms elsewhere.

The analysis of an electromagnetic design by the finite element technique and the subsequent result retrieval can be divided into several sections: preprocessing, solving, and postprocessing. In principle, these amount to setting up the problem for solution, carrying out the mathematical work, and evaluating the answers to produce some useful result.

Preprocessing

Preprocessing is the working phase in which the problem to be solved is described to the computer. A complete problem definition consists of two parts, which may be separated. The first is the geometric and topological description; the device has a certain shape and is made of certain materials which must be described to the CAD system. If analysis is regarded as the simulation of experiments, then this part of problem definition may be viewed as simulating the construction of a prototype. In general, the prototype is modelled by a set of finite elements in analysis, and most current systems combine a description of the device itself with a statement of how it is to be represented in terms of finite elements. The second part of problem definition may be regarded as describing the experiment to be simulated on a prototype device; it includes boundary conditions and excitations. Both of these parts may be input to the computer by the use of an interactive program in which the user is, essentially, editing an existing data base. The provision of a local workstation with graphic displays and input of the type described above can make this part of the analysis process both simpler and more natural for the user than the current manual methods. The final product of preprocessing is a completely defined, unique, mathematical problem ready for solution. Consequently, the ultimate goal of a preprocessor is well defined.

Solution

Preprocessors are really problem-defining programs, which arrange for all necessary physical data to be assembled. Solvers are programs for constructing the systems of algebraic equations which model the physical situation mathematically and produce solutions. Solvers are very frequently set up as batch processes, requiring little if any interactive control by the designer. They can therefore be implemented on large remote computers as an alternative to the local workstation. In most magnetics problems, and in nearly all currently available packaged systems, the end product of the solver is a set of potentials describing the magnetic field in the entire model.

Postprocessing

The potential solution produced by the solver provides the raw information about the electromagnetic field within the device, subject to the approximations made in the modelling and solution phases. These data are not necessarily, indeed not usually, what is required by the designer, whose needs are much more likely to include derived quantities like impedances, power densities, or mechanical forces. However, the required results may be extracted from the potential solution by further mathematical manipulation. The postprocessing facilities of a CAD system therefore must provide two main functional ingredients: mathematical manipulations for treating the available solution data and graphical tools for displaying the results thus obtained.

In contrast to the preprocessing phase, it is difficult to predict in other than broad terms what information might be needed as final output from a postprocessor. Typical output might include flux plots, energy and inductance calculations, terminal parameter evaluation, loss calculations, local field values, and so on; but almost every design task imposes requirements not encountered elsewhere. If computational analysis is viewed again as a simulation of experiments, then postprocessing amounts to deciding what instrumentation should be provided and what results should be recorded. Postprocessing therefore needs to be done by system modules with a great deal of flexibility, so that the many and varied needs of the day can be accommodated.

Practical Examples

Throughout this book, theoretical discussions are illustrated by practical examples. Most of the examples are brought from one particular CAD system, *MagNet*, almost all the remainder from two others, *MAGGY* and *PE2D*. Of course, there are numerous other systems, with varying characteristics; indeed, a later chapter of this book surveys them in some detail. But although their working characteristics differ widely, the capabilities of the various systems are functionally very much alike. They differ, as discussed elsewhere in this book, in their ease of use and in their adaptability to various hardware and software environments. Consequently, the design and modelling techniques described here may be applied, *mutatis mutandis*, to the various other systems as well.

Magnetic Material Representation

Analysis of magnetic devices normally requires a knowledge of the physical properties of the materials used: their magnetization curves as well as various other characteristics. Computer-readable files of material properties therefore must be maintained ready for use as and when required, in a convenient form. Ideally, the manufacturers' catalogues for all materials to be used in design should be computer-readable and computer-accessible, so that analysis programs need only refer to materials by name or type identification.

CAD systems differ widely in the level of detail and the precision of modelling used for magnetic material characteristics. Furthermore, systems differ in how the files are arranged, how data can be entered, and what file management facilities are available. The overview given in this chapter, though far from exhaustive, considers both the mathematical modelling and the file maintenance aspects of the problem of material property representation.

Modelling of *B-H* Curves

Numerous ways of modelling magnetic material property curves have been devised. The task is surprisingly difficult, in view of the relatively simple mathematical nature of *B-H* curves. Most mean magnetization curves are after all single-valued, continuous, and monotonic, often with at least the first derivative monotonic also.

Nature and Description of *B-H* Curves

Even if such complications as hysteresis and anisotropy are ignored, *B-H* curve modelling is still not a trivial task. Difficulties arise from several sources. First, the range of slopes encountered (the range of incremental permeabilities) is very wide in ordinary as well as esoteric materials. The steepest and flattest slopes for a single curve may differ by a factor as large as 10^4, straining the precision limits of the conventional 32-bit computer word which retains about 7 (decimal) significant figures. Secondly, precision difficulties in curve modelling are merely a mathematical image of the difficulties encountered in measuring the characteristics in the first place. Data are usually obtained by measuring the flux density B that corresponds to a given current, hence to a given field H. But designers often need to know the field H that corresponds to a given flux density B, not the other way around. For saturated materials, that implies accuracy requirements bordering on the extreme, because the magnetization curves of most ordinary materials are nearly horizontal at high saturation levels, as illustrated in Fig. 1. It should be noted that no attempt has been made here to model the so-called *Rayleigh region*, that portion of the mean magnetization curve very near the origin where the curve is concave upward.

The equation solving techniques currently popular do not restrict the magnetic material characteristic unduly. However, they do use the slope of the curve, as well as curve values, in computation, and they often converge faster if the slope is monotonic. Furthermore, they generally proceed from flux densities to fields and therefore favor the storage of magnetic characteristics in the form of a curve giving the material reluctivity ν (the inverse of permeability) as a function of the squared flux density B^2. However, some analysis systems store the B-H curve in its classi-

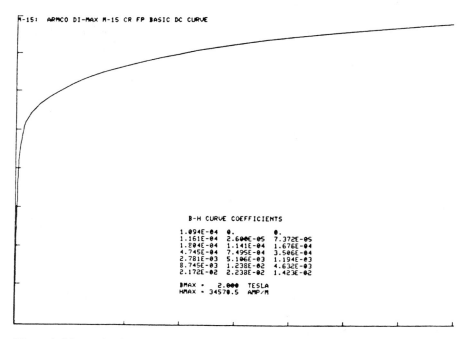

Figure 1. Magnetization curve of an electrical sheet steel. The curve is very nearly flat at high saturation levels, leading to problems of modelling precision.

cal representation, $H = H(B)$, then calculate the reluctivity and its derivative as needed. In some other systems, particularly those aimed at scalar potential solutions, the permeability is stored as a function of the magnetic field H. Thus, the most popular curve models now in use by CAD systems are, in order of their apparent popularity,

1. Reluctivity as a function of flux density squared, $\nu = \nu(B^2)$.
2. Field as a function of flux density, $H = H(B)$.
3. Permeability as a function of field squared, $\mu = \mu(H^2)$.

There are two good reasons for using squared values of the independent variables B or H, rather than magnitudes. First, these variables are usually derived from potentials in vector component form, so that finding the magnitude involves first finding the squares of the components and then extracting the square root of their sum. The relatively expensive square root operation is avoided and work saved by tabulating in terms of the squared value in the first place. A second, and perhaps more important, consideration is numerical stability. Working with the square tends to emphasize curve behavior at high flux densities or fields, where higher precision is usually required.

Curve representations in early CAD systems were most often chosen to minimize computing time. With modern equipment, the cost of a few extra multiplications is becoming more or less insignificant. The choice of representation is therefore dictated increasingly by the *robustness* of representations, that is, by the range of unusually demanding cases that can be handled without likelihood of failure. A second important criterion is user convenience in filing, editing, and data file maintenance. It takes second place not because it is less important, but because flaws in convenience can often be overcome by clever design of the user interface and clever programming—a lack of robustness cannot.

Internal Form of *B-H* Curves

No matter what mathematical modelling technique may be employed to represent *B-H* curves, the very nature of digital computers requires data to be stored in tabular or list form, as sets of numbers. As a result, curves are always filed as sets of numerical parameters, usually the coefficients of certain approximating functions chosen in advance. As a simple, familiar (but not often used) possibility, curves could be represented by Fourier series; all that need be stored in that case would be the coefficients of the various trigonometric terms in the series. The functions themselves need not be stored, since they are always the same.

Most mathematical methods now in use are able to reproduce curve data with greater precision than is warranted by the precision with which *B-H* curves can be determined in the first place. Thus, curves known in one of the various possible representations can usually be converted into another form without significant loss of accuracy. In every representation now in use, the approximated quantity $a(x)$, which may be the reluctivity or permeability, is given as a function of an independent variable x, typically the squared flux density B^2, in the form

$$a(x) = \sum_{i=1}^{N} q_i f_i(x).$$ (1)

The functions $f_i(x)$, $i = 1, \ldots, N$, are members of some reasonable family chosen in advance. These may include functions that bear a close resemblance to the curve being modelled (e.g., exponential functions), polynomials of one sort or another, or they may even be simple piecewise straight line segments. Typically, the number N of such functions ranges from about three or four, up to about 50. The larger numbers are generally associated with the simpler approximating functions. As an example, the MagNet system uses a set of 14 cubic Hermite polynomials, which guarantee continuity of both the function $a(x)$ and its first derivative.

The internal form of representation is not actually of much importance to the user, except possibly when things go wrong. When in trouble, informed guesswork based on an understanding of internal system procedures can be helpful. For example, it may happen that a proposed device design involves flux densities far beyond those permitted by the available B-H curve. In manual design, the usual cure is to extrapolate the available curve and hope for the best. In computer-aided design, life can be more difficult: it may not be obvious that the curve range has been exceeded, nor may it be evident what sort of extrapolation is used by the particular CAD system. There are no extrapolation methods guaranteed to work every time, and an understanding of the mathematical methods underlying the programs may be helpful to the user when results begin to look suspicious.

A vital part of the information stored for any material characteristic curve is a short descriptive name, an additional legend or identifying text, and a statement of the units in which the data were first entered into the file. These added details make the machine-readable file mimic the data sheets given in the usual printed materials catalogue. After all, a curve is of no use in splendid numerical isolation; it is necessary to state the units in which its axes are calibrated and to accompany it with a legend or title which identifies it. A useful legend is generally far more extensive than can be accommodated in a computer file name of six or a dozen characters. Well designed CAD systems should make adequate provision for including such textual information; it is not good enough to tell users to choose clever file names within their own computer systems!

Management of Material Property Files

The storage of magnetic material property files, containing B-H curves and other characteristics, in computer-readable form is relatively undemanding compared to, say, files containing geometric shape information. However, the management problems encountered in maintaining files and providing easy access to them are very similar regardless of file content. The filing and editing of magnetic property curves will therefore be dealt with both for its own sake and because it is one area in which it is easy to concentrate on the data handling aspects without becoming sidetracked in mathematical or computational issues.

File Characteristics

Files used in computer-aided design of electromagnetic devices may be classified as working (temporary) or archival. Working files are of a scratchpad nature: here today, gone tomorrow, and good riddance. Material properties, on the other hand, constitute information of archival value, so it is appropriate to keep material data in files which serve no other purpose. In particular, it is best to keep them entirely separate from specific design problems to be solved, making working copies as and when required.

From a CAD system user's point of view, the significant questions to be considered are:

1. What do the material property files contain, how are such files managed, and what must be known to use them?
2. How can data be entered or modified in such files?
3. What mathematical methods are used for curve data encoding?

For most ordinary user purposes, the third question may seem more or less irrelevant. For CAD system evaluation, however, it is worth asking because the choice of mathematical technique can affect numerical stability, accuracy, and computing time. Furthermore, the choice of mathematical method has a significant influence on file size and organization and may be of some interest on that account.

Representation of a Single Material

In a manufacturer's catalogue, each material is likely to be described and characterized by several curves. There is always a mean magnetization curve, ordinarily a loss curve or two, and perhaps others.

Archival files of magnetic material properties properly should contain any and all information available about each material. It is not useful to assert that only a restricted variety of information, such as the mean magnetization curve, will suffice. The experience of CAD system designers shows that no matter what reasonable assumptions are made about the desires and needs of users, a maverick user with different requirements will turn up soon enough. However, it is safe to assume that a mean magnetization curve will always be required. For the rest, one cannot do better in file design than to budget for the computational equivalent of blank graph paper, that is, to provide for a number of material property curves of types not specified in advance.

In the MagNet system, which will be referred to frequently in this book as an example of a state-of-the-art CAD system, each material is assigned four "graph sheets", one of which must always represent the mean magnetization curve (B-H curve) of the material. The other three curves may be used to model any desired single-valued property. Examples might include loss at 60 Hz, incremental minor loop permeability, and magnetostriction coefficient. The mean magnetization curve is considered fundamental—no material can be made to exist without one—and the maximum value of flux density that occurs in the mean magnetization (B-H) curve is held to be the maximum value for all auxiliary curves as well. Each of the four curves is represented by a similar set of approximating function coefficients, and each is accompanied by a descriptive legend or text as well as a statement of the units used.

The data structure that represents a single material has the form shown in Fig. 2. The *material* itself is a small file segment which comprises a descriptive text to identify the material, names of the four (or fewer) curves that go with the material, and pointers to the storage locations of the curves. Each curve in turn comprises the actual numerical data (curve coefficients), as well as information on units, and a legend that describes the curve.

Material Libraries

Collections of materials are dealt with by combining them in *libraries* of materials. For ease of subsequent identification, every material can be given a name before it is filed in a library and can then be retrieved by name. Material names are of course distinct from the curve names assigned to each curve associated with a given material. A hierarchical naming system thus results, in which a particular curve is identified by both the material name and its own curve name. For example, IRON LOSS might identify a curve named LOSS, associated with a material called IRON.

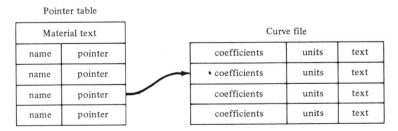

Figure 2. Curves describing material properties are filed in a common file space. They are retrieved by name, by looking up the storage locations in a table.

A file containing material properties must of course be equipped with its own file directory, so that the various materials can be found in it as and when required. Therefore, a library of materials must consist of a file containing the material descriptions and a directory which identifies the contents of the file. In practice, the directory and data file are stored in a single computer file; but such internal details do not normally concern the system user.

The directory contains a set of material names, in which each name carries with it a pointer to show where in the file itself the corresponding material description is to be found. In other words, the directory is a reference table of names and file locations. The file proper is organized as a set of materials, each comprising at least one curve. These requirements lead to the hierarchical data structure shown in Fig. 3.

How many curves to permit for each material, how many coefficients per curve to store, and what limitations to apply to the identifying text, are decisions the system designer must take. Some of the currently available CAD systems permit only one curve (the *B-H* curve) to be given for each material; others are more generous. Ideally, of course, the number of curves should be unrestricted except by total file capacity.

File Management

Most designers deal only with a relatively small number of materials in their everyday activity, typically not more than five or ten and often only three or four. But for each of these, several computational models may be required. For example, curves may be given for the upper and lower decile production tolerance limits of a material. Similarly, it may be desirable at times to have more than a single characteristic curve available, the

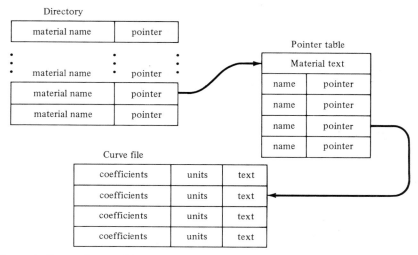

Figure 3. Curves form a hierarchical data structure, in which the library directory contains materials, and materials in turn possess curves.

several curves being adjusted for particularly good accuracy at very high or very low flux densities. The number of *B-H* curves for any one physical material may then reach half a dozen, and the apparent number of materials (so far as the CAD system is concerned) may well be two, three, or even ten times larger than the nominal number of physical materials. The requirements of a design office, rather than just a single individual working on a particular project, may quite easily approach fifty or a hundred. For convenience in design work, it should be possible to house this number of materials in a single materials library, so that all materials a designer might need become equally accessible.

Whether a single large materials library file should be maintained, containing all conceivable materials ever used by a design office, or whether it would be wiser to have individual engineers or working groups establish smaller data bases of their own, is very much a matter of taste. The former approach has the merit of making everything accessible and known to everybody. On the other hand, chaos will ensue if every member of a large design group is permitted to alter the library at will. An extensive single library therefore entails creating a library management function and a mini-bureaucracy (sometimes not so *mini!*) to go with it. A collection of many small libraries may be easier to use, but it can lead to work duplication in a large organization, as several people independently discover the same data.

Rarely is the materials library of a design office static. Even though no major changes may take place for years on end, it is reasonable to expect that alterations and updates of a minor sort will occur frequently, even daily. New materials may need to be inserted or old ones deleted; new curves (e.g., valid for a higher maximum flux density) may replace existing ones; auxiliary curves may need to be replaced with new ones, and so on. Thus, it is essential to provide for a range of operations which may be termed *file management*. It is desirable, to say the least, for the CAD system to allow the user to do the following:

1. Display a catalogue of all materials now in the library.
2. List the curves on file for a particular material.
3. Delete one or more materials.
4. Delete any one curve belonging to a material.
5. Make a copy of any one material, with a different name.
6. Rename an existing material or curve.
7. Scale the axes (e.g., to allow for stacking factors).
8. Plot any desired curves for inspection.
9. Extract materials or curves in an externally useful form.
10. Merge two distinct libraries.

In addition, it is essential to possess the following capabilities:

11. Modify the describing texts that go with the curves.
12. Create new materials and curves.
13. Modify the material properties themselves.

Only the last two functions really have anything to do with the mathematics of curve modelling; the other eleven are concerned with file management and are equally applicable no matter what form of representation is employed for the curves themselves. This situation is typical of CAD systems generally, and of the MagNet system in particular: only a small amount of the visible system structure (and a small part of the program code) are actually concerned with the mathematical techniques. The editing functions, important as they are, occupy only a minor portion of the system. The remainder is concerned with file management.

Security and Privacy

In any reasonable working environment, security arrangements must also be acceptable. Files must be protected both in the physical and legal senses, i.e., against computer failure and human error as well as against mischief and evil intent. Physical protection is most easily provided by copying files onto a medium which may be removed from the computer, such as magnetic tape or floppy disk. Protection in the legal sense is easy to achieve in single-user working situations where the computer and all file storage devices are under the direct control of the designer. It is much more difficult to obtain in multi-user systems where disk space and computer time are shared with other users. Clever schoolboys equipped with telephone and terminal have been known to penetrate supposedly foolproof password schemes. Perhaps the best security scheme is to make copies on removable media and to leave as few files as possible on the public storage device.

Curiously enough, computer-readable files probably satisfied all these requirements best during the punch-card era. Indeed, the punched card is a unique medium, being both machine-readable and man-readable. A great deal of file management which now appears very sophisticated was carried out in that long-ago age by simply shuffling decks of cards. Magnetic media are unfortunately readable by machines only, so that even the most trivial operations have to be carried out by programs designed for the task. Since cards are easily portable and removable from the computer, security arrangements were not difficult to make. Data privacy and security are usually well served if users are conscious of the need for security and are able to remove files physically from the computer system. Floppy disks and cartridge tapes are excellent media from that point of view!

Data Entry and Editing

Every file used by a CAD system must be capable of modification as well as initial creation. Every type of file associated with a CAD system must therefore be associated with at least one *editor*, a program that renders the user able to insert new materials into the file or to modify the existing file contents.

Editors

Creation and modification of curves is generally the task of a curve editing program. Because it is desirable to have the editor communicate with
the user in ways that make sense in terms of electromagnetics, not in
terms of file structures, a special editor is required for material property
files. A different editor will be needed for geometric shape files, and so
on—one editor for every file type.

Although material property files may be regarded as archival in nature,
users frequently need to define new materials. Quite often these may not
be physically real at all. For example, laminated magnetic materials are
often numerically simulated by scaling their B-H curves in accordance
with some numeric stacking factor. Such curves have little permanent
value, except perhaps to record the assumptions on which a particular
design was based. Since unusual materials, intended to be employed in
one or a few projects, do also occur, easy entry of new curves into the
materials library is important.

Editors are invariably interactive programs. Where data are essentially
graphic in nature, as in almost all CAD situations, an editor must incorporate its own form of graphic display and must be equipped with its own
command language. Display requirements for curve editing are straightforward in principle—in essence, it is required to reproduce the conventional curve sheets in manufacturers' catalogues. But the range of display
options can cover a wide range in practice, since both logarithmic and
linear axes, often with suppressed origin, are routinely used. Furthermore, units are quite varied and are often mixed; for example, B-H curves
are on occasion encountered with B given in kilogauss and H in amperes
per inch!

The command language of a material curve editor may take different
forms, but it is desirable for it to include at least some graphic input facilities. Of the conventional hardware devices for graphic input, only the
graphics tablet (digitizer tablet) is useful for communicating numeric
information. Other devices—joysticks, thumbwheels, lightpens, trackballs,
mice, touch-screens—are useful for screen pointing but cannot transmit
positional information with accuracy and resolution comparable to that of
ordinary graph paper. Even a tablet is not useful unless accompanied by
some terminal intelligence, and suitable software or firmware to carry out
certain minimally necessary functions such as cursor tracking. If a tablet is
not available, or not suitably supported by the computer system, input
operations must proceed entirely through numeric (keyboard) operations.

Curve Creation and Editing

When a new material or a new curve is created, all components of the file
entry that corresponds to it must be created at once, if a valid file entry is
to exist. There are two ways of satisfying this requirement: by locking the
user into a sequence of actions which only terminates when every

required data item has been entered, or else by assuming default values, and then allowing the user to alter any desired items. Although there exist practical CAD systems which lock the user into a forced dialogue, neither extreme is inherently desirable. A compromise is therefore indicated.

In the MagNet system, the compromise is adopted of asking the user to specify two things: the identifying text for the curve to be created, and at least one point on the curve. For new materials, the maximum value of flux density is required for which the description is to hold, so that if only one point on the curve is given, it must be the point corresponding to the maximum value of B. This requirement is necessary for graphic echoing of input; if the maximum value of B is not known, it is not possible to choose axes and scales to accommodate the curve. If no other points are entered, the curve is then assumed to be a straight line passing through the origin. If this assumption is not correct, the user may of course edit the curve to correct it.

Creation of new materials is handled internally in MagNet as if it were an editing operation: when a material is declared to exist it is automatically assigned a vacuous set of properties, and all newly entered points are regarded as part of an editing operation. Thus, identical operation sequences, command syntax, and graphic techniques apply to editing an established curve or creating a new one.

At any given time, a material property library may contain dozens of different materials. When creation, editing, or indeed any other operations are undertaken on a particular material, that one is retrieved from the library (or created, if it did not yet exist) and declared to be the *active material*. Until the user switches to another material, all operations are taken to relate to the active material. This way of working closely resembles the accustomed handling of paper records: one curve sheet or material file is selected from the entire library, and any following sequence of operations is understood to apply to that particular one.

Magnetic quantities are measured and recorded in a bewildering array of units, from gilberts to Kapp-lines. Chaos is best prevented in CAD systems by separating the units used externally (the *user's units*) from those employed for internal storage. For example, MagNet stores all quantities in SI units. However, the user is normally unaware of the internal units and indeed has no reason to know what they are, since the user units are recorded in the materials library. Data may therefore be entered in megamaxwells per acre and abamperes per furlong, if the user so wishes. Although the curve will be stored in SI units, the original user units will be remembered and the curve will be reproduced in these units if desired.

Numeric Data Entry

Magnetic material data are normally obtained in the first instance either in the form of graph sheets published by the material manufacturer or as a numeric tabulation. Numeric tables may reach the user recorded on paper, or they may already reside in a computer-readable file. Ideally, both forms should be catered for by CAD systems.

Data stored in computer-readable files are perhaps the easiest to treat, even if the file format is a little inconvenient. For example, the tabulation may be arranged randomly, or it may present ordinate and abscissa values in inverted order relative to the system's expectations. It may then be necessary to write and use separate small programs to convert the data file into a suitable form. Computers are rarely able to guess which number in a pair represents the flux density and which one represents the loss!

Numeric data entry is encountered in two forms: the so-called *conversational* form and *quasi-batch* input. In the former, the system prompts the user, giving the value of B (or H) and expects the corresponding value of H (or B) to be entered. The *quasi-batch* form of entry is more open-ended and permits the user to give a sequence of points on the curve, without any particular restriction as to their placement along the curve. The latter is clearly preferable in terms of convenience; it allows, for example, for direct entry of experimental points without any need for interpolation.

When numbers are entered, they may be supplied either as a file or through the keyboard. Keyboard entry is perhaps the least convenient form of data entry for most users. It is slow and error-prone, and it is quite difficult to check for mistakes afterward. A better working procedure is to employ a standard text editing program to create a computer-readable file, which can be verified and any errors corrected. That file can then be read just like any other. An advantage of this approach is that the original data file remains available for re-use, should it be required.

Graphic Data Entry

Graphic entry of published data is the natural procedure when the data are presented in graphical form in the first place. Using a digitizer tablet, whose spatial resolution is typically around 0.1 mm, input will be as accurate as it is possible for human eye and hand to be. Higher requirements cannot be placed on published data sheets in any case, since they are subject to at least that much error by virtue of the dimensional instability (including nonlinearity and anisotropy) of graph paper, and contain printed lines of finite width.

The graphics tablet resembles a largish pad of paper, about 40 cm square and 15 mm thick, on which the user can "write" with a pencil-like stylus, which actually steers a crosshair cursor on the screen. The cursor usually appears as a movable crosshair, arrow, or other aiming device, as illustrated by Fig. 4. The writing motion seems very natural to most people, and hand–eye coordination is as good as it would be on paper. An alternative input device employed with digitizer tablets is the puck. Pucks are flat objects ordinarily equipped with a plastic window or lens, with engraved crosshairs. For most graphic operations, users generally like a

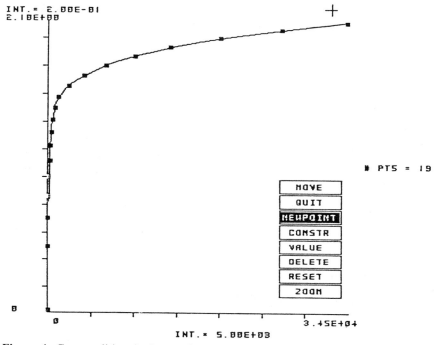

Figure 4. Curve editing in interactive work involves choosing editing functions from a menu and identifying points, using a crosshair cursor as a pointing tool.

stylus; for precision in graphic input, the puck is preferred by some. As indicated in detail elsewhere in this book, many other common graphic input devices, notably mice, lightpens, and joysticks, are unsuitable for digital encoding of graphs, they are only useful for freehand sketching.

Graphics work in the MagNet system, and in practically all other systems as well, employs either two separate screens, or a single screen partitioned into two or more distinct areas usually referred to as *windows*. One normally carries on the conversation with the keyboard, another is attached to the graphics tablet. All graphic input operations are accompanied by menus on the screen, showing what graphic actions are currently available. Pointing is done by steering a crosshair cursor with the graphic input stylus. Thus there are at least two channels of communication between the system and the user—keyboard and tablet. Their use is kept strictly orthogonal, in the sense that whenever a menu is displayed and the crosshair cursor is presented on the screen, graphic input is expected. Keyboard input is not accepted at that time. Conversely, when no cursor and no menu appear in the display, a user prompt appears in the window devoted to keyboard conversation, showing that only keyboard input will be accepted at that time. In some systems, these distinct display windows in fact use the same screen space; one or the other is visible at any one moment, and the user can switch back and forth between them. The exact arrangements depend principally on the type of computer display hardware in use.

It has been found experimentally that most human operators are relatively inaccurate in attempting to follow a curve smoothly with puck or stylus, but much more accurate when asked to identify individual points on a curve by means of the same device. Thus, the reasonable method for entry of curves is once again through the medium of a set of sampled points. Working with a graphics tablet, however, points can be established quickly and pleasantly; it takes most users a few minutes, or less, to choose and transmit 30–50 points well spread on a *B-H* curve.

Fundamental problems in graphic input concern setup and calibration of the graph sheet. Graph paper comes with various sorts of axes and grids; furthermore, it is not sufficiently stable dimensionally to have the two axes orthogonal to within the accuracy of a normal graphics tablet. Placement of a graph sheet at a specified position on the tablet and alignment of paper axes with the tablet axes are impossible tasks in practice. Any graphic input work therefore must begin with a setup procedure, in which the placement and scaling of the graph is communicated to the system. A typical setup procedure requires the user to fasten the paper to the tablet and then to inform the system about graph placement, scaling, and calibration in an interactive conversation. That query-and-answer sequence requests the user to identify, by means of puck or stylus, the origin and the extreme calibration marks on the graphic axes, so as to identify paper location and orientation. When the calibration procedure is complete, an image of the axes is produced on the display screen, thereby giving the user an opportunity to spot any obvious mistakes.

An essential ingredient of interactive computing is a knowledge on the user's part of just exactly what is going on, and just exactly what has happened to the data. To this end, graphic echoing is the most important tool available.

Ideally, the graphic screen image should provide the user a detailed, and fairly complete, representation of the data base in use. There are many possible representations of a given data set—for example, a set of points along a *B-H* curve could be represented as a *B-H* curve in the conventional way, or as a curve of permeability against flux density. The plot could be on one of several kinds of graph sheet, possibly with suppressed axes; it could show the curve itself, or the sample points which underlie it, or both. For echoing graphic input activity, representations are generally chosen which correspond to the kind employed by the user, so as to make the tablet and screen as nearly similar as possible.

Graphic input is preferable to keyboard work not only because it eliminates tiresome numeric typing, but because it allows editing with greater ease. In particular, deletion of a point (in case of error, for example) is a complicated matter in keyboard work. In graphic work it is quite straightforward to provide the user with commands which say, in effect, "delete this one"; the point to be deleted is identified by pointing at it with the graphic input device. However, data entry through the keyboard must always remain, for the keyboard is the only known device which will easily allow specifying data with six-figure precision!

Curve Editing Operations

Editing operations may be required for a variety of reasons. Additional data obtained about a material may make it desirable to correct the existing curves, or the numerical curve modelling process may have produced a poor fit. Poor fits may result, for example, if too many or too few points were used to describe some particular part of a curve, so that the data in that region were weighted too heavily or not heavily enough compared to other parts of the curve. Such flaws are often quite obvious on visual inspection and may be corrected by entering additional points on the curve.

Curve editing is carried out in MagNet very much like curve construction in the first place. Individual data points are entered or deleted as required, and the curve is then reconstructed out of the new point set.

Magnetization curves as retained in a material file are true continuous curves, since they are stored as parameters associated with approximating functions selected in advance. To edit a curve, it is first recast as a set of data points, which are so chosen that the particular curve fitting method employed will reconstruct the original curve if no alterations are made. As in initial data entry, new points may be entered from a file, through the keyboard, or by way of the graphics tablet. In the latter case, a setup and calibration procedure will of course be necessary, so that the system can tell where on the tablet the graph sheet is located, what its orientation is, and how its axes are calibrated.

The similarity of editing operations to initial curve building is not coincidental. In fact, the initial construction of a curve is handled by the editing program: when a material is declared to exist it is automatically assigned a vacuous set of properties (zero points prescribed on the curve), and all newly input points are regarded as part of an editing operation. Thus, the same changes of mind that apply to editing automatically apply to data input.

In a true interactive editor, the representation of material properties involves exactly one file, and all operations are carried out on that file alone. There is no distinction between "unedited" and "edited" files, and no distinction between continuous-function representations as against point sets. The price paid for unity is that retention of the "original" point set is not practical. The benefit gained is that any operation of which the system is capable can be performed on any file at any time. Contrary to common practice in batch-run programs, there are no "phases" or "modes" which follow one another in sequence; there is only an editing mode, in which any data can be altered and corrected. Second thoughts and error corrections are thus possible at any time. On the other hand, the organization of programs in sequential "phases" implies that once a "phase" has been terminated no further error correction is possible.

Batch Operation

In contrast to interactive systems, batch-run programs expect to read their input data all at one time and, perhaps more significantly, they expect all data to be in the final, correct form which the user intended. There is no provision, or at the most only rudimentary facilities, for any changes of mind or any correction of errors, once the computation has started.

Whether it accomplishes its work of analyzing magnetic fields on an interactive or a batch-run basis, any CAD system capable of maintaining a library of material property files must provide the basic file management functions. Searching, cataloguing, retrieval, copying, and renaming of materials must be possible, and some form of display operations must exist. However, these need not necessarily even be a part of the CAD system itself. There are useful and good CAD systems which maintain their material curve data simply as tabulations of B and H values, then recompute all derived quantities at every program run. The advantage of doing so, in environments which permit interactive batch entry to the computer operating system (as distinct from interactive CAD), is that the file management programs, text editors, and other utilities provided by the operating system can be used for management of material libraries as well. This scheme restricts all data handling to be done on a numeric basis, not through graphic operations.

Complex Magnetic Materials

Many practical magnetics problems involve magnetic materials less simple than the classical narrow-loop substances commonly referred to as *soft*— for example, permanent magnets or anisotropic materials. Quite a few present-day CAD systems are equipped to deal with remanence, or anisotropy, or in some cases with both. True hysteretic materials, by contrast, can be handled only in a few very restricted cases. Because the physical mechanisms of remanence and hysteresis are complex and dependent on material history, the mathematical models now provided in CAD systems tend to be empirical, chosen to reproduce macroscopic behavior well over a restricted range of problems. They are generally not based on descriptions of the actual material physics.

Permanent Magnet Materials

A full description of hysteretic phenomena, particularly of minor loops, is not even attempted by presently available CAD systems. The usual treatment of magnetically hard materials consists of shifting the known demagnetization curve (the first and second quadrants of the *B-H* plane), so that the demagnetization curve passes through the origin, and recording the amount of the shift either in terms of the coercive force H_c or the remanent flux density B_r. For analysis purposes, the material is then treated as if it were soft, and the coercive force or remanence is accounted for by introducing fictitious distributions of poles or currents equivalent to the coercivity or remanence. Usually, the material is assumed to possess a magnetization characteristic symmetric with respect to the origin, i.e., its shape in the third quadrant is simply assumed to be a mirror image of the first quadrant of the *B-H* plane. Such a shift is illustrated in Fig. 5.

The simplest technique for treating permanent magnet materials, available even where the CAD system has no knowledge of remanence, is to shift the curve as in Fig. 5, then to supply the coercive force H_c by introducing into the problem suitable current-carrying coils. Where permanent magnet materials have simple shapes, such replacement of remanence by equivalent coils is straightforward, and any convenient analysis system capable of dealing with soft materials can be pressed into service using this approach. Analysis systems which do make provision for hard materials, on the other hand, merely require the user to specify H_c, or perhaps B_r. They then modify the field equations internally so as to create the distributed equivalent of current-carrying coils. This internal treatment is of course externally invisible, so the system user has no need to be concerned about its details.

True magnetically hard materials are usually anisotropic, so they should be characterized by a tensor rather than a scalar permeability or reluctivity. A simple treatment adequate for many real materials consists of taking the reluctivity (inverse permeability) to be given by the matrix form

$$\nu(B) = \frac{1}{\mu_0} \begin{bmatrix} \nu_r(B_p) & 0 \\ 0 & c \end{bmatrix} \tag{2}$$

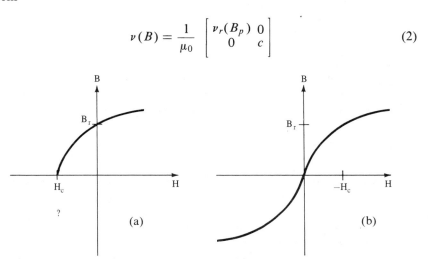

Figure 5. (a) Demagnetization curve of a hard material. The curve is unknown in the lower half plane. (b) The shifted curve used in analysis.

The geometric axes are assumed to be so chosen that the principal axis, along which B_p is measured, coincides with the direction of the remanent flux density. The constant c is a characteristic of the material; its value is

$$c = \frac{\mu_0 H_A}{\mu_0 H_A + M_S} \tag{3}$$

where H_A is the *anisotropy field* and M_S represents the *saturation magnetization*. Both are descriptive parameters often available from the material maker as material specifications. In the case of rare-earth materials such as samarium cobalt, the constant c is very close to unity. Thus, a reasonable model is obtained by simply taking $c = 1$, which amounts to saying that the material acts in the crosswise direction as if it were air.

The techniques of material characterization used in current systems are empirical and represent crude approximations in some cases. Surprisingly, they are very accurate for most of the newer magnetic materials, such as samarium cobalt, because many rare-earth materials have almost straight demagnetization curves. Although accuracy tends to be lower when dealing with some conventional materials, it usually suffices for design purposes.

Geometric Anisotropy

Where laminated magnetic materials are used, the laminations are usually not modelled separately. Instead, they are aggregated into a homogeneous composite material which is given a hypothetical magnetic characteristic. The common way of doing so is to treat the material as anisotropic, with the principal axes parallel to and across the lamination planes. The reluctivity tensor then takes a form similar to that given above for magnetically hard materials. Subsequent mathematical treatment follows that of the permanent magnet material, except of course for the value of the remanent field, which must be zero.

Suppose magnetic material laminations of thickness w_i are separated by interlamination air spaces of thickness w_a. In general, the flux density B will be at some angle θ to the laminations, so it may be resolved into components B_x and B_y,

$$B = \mathbf{1}_x B_x + \mathbf{1}_y B_y \tag{4}$$

as shown schematically in Fig. 6. An approximate model of this material can be obtained by assuming that it is adequately representable by a diagonal tensor, whose two diagonal components are such as to give the correct B-H relationship for flux-line incidences in the plane of the lamination and normal to the lamination.

Suppose first that all flux lines are in the plane of the laminations. In general, there will also be some flux in the interlamination air space,

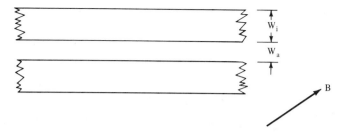

Figure 6. Iron laminations with interlamination air gaps, subject to an arbitrarily directed flux density.

because the air space and iron are subjected to the same magnetic field H_x. Thus

$$B_x = \mu_i H_x \tag{5}$$

in the iron, which is assumed to have a permeability of μ_i, and

$$B_x = \mu_o H_x \tag{6}$$

in the air. The amount of flux there will be smaller than in the iron, but it may nevertheless be significant, especially in highly saturated cases. In one lamination and one air space, the total flux per unit depth is therefore

$$\phi = w_i \mu_i H_x + w_a \mu_o H_x. \tag{7}$$

The flux density, averaged over the iron and air, is consequently related to the average magnetic field by

$$B_{ave} = [s(\mu_i - \mu_o) + \mu_o]H_{ave}, \tag{8}$$

where s is the effective *stacking factor*,

$$s = \frac{w_i}{w_i + w_a}. \tag{9}$$

The stacking factor is a pure number always smaller than unity. In effect, equation (8) says that more field H_{ave} is needed for a given average flux density B_{ave} than would be the case in a homogeneous slab of iron.

Consider next the opposite extreme, in which the flux density is directed normally to the laminations. The flux density B_{ave} is now the same in iron and air; the field H, on the other hand, varies considerably. The magneto-motive force across one lamination and one air space is clearly

$$\int_0^{w_i + w_a} \mathbf{H} \cdot \mathbf{dl} = \frac{w_i B_y}{\mu_i} + \frac{w_a B_y}{\mu_o}. \tag{10}$$

Solving for $B_{ave} = B_y$,

$$B_{ave} = \frac{\mu_i \mu_o}{\mu_i - s(\mu_i - \mu_o)} H_{ave}. \qquad (11)$$

In other words, a given field H will again produce less flux than it would in solid iron, but less by a factor quite different from that in the previous case. As would be expected, the iron predominates when flux lies in the plane of the laminations; the air is the governing influence with flux density normal to the laminations.

Equations (8) and (11) give values for the diagonal elements (*xx* and *yy*, respectively) of the permeability or reluctivity tensor, whichever may be desired. The off-diagonal elements are assumed to be zero for simplicity. The resulting material description is only an approximation, but in most ordinary materials it appears to serve rather well for design purposes.

Curve Modelling Theory

At first glance, the modelling and editing of characteristic curves may appear to be almost trivial tasks, yet on closer examination they turn out to involve both substantial mathematical content and quite a bit of computational complexity. Neither is of primary interest to the CAD system user. On the other hand, both may repay brief examination. An understanding of the mathematical methods employed can help the sophisticated user make the most of any CAD system. This section consequently gives a brief overview of the main theoretical issues; it is intended for readers acquainted with computational mathematics and may safely be omitted on first reading.

Curve Representation

There now exists an extensive literature on the choice of mathematical functions for modelling the magnetic characteristics of soft iron, both in the form of approximations to B-H curves and to reluctivity curves. Polynomial, exponential, rational algebraic, and various other functions have been employed to this purpose. Hermite polynomials and cubic splines are fairly popular choices. In general, it would appear that almost any reasonable set of ten or twenty linearly independent continuous functions with continuous first derivatives can represent magnetization characteristics to the precision warranted by the precision available in measurement on the one hand and the assumptions regarding uniqueness on the other. Criticisms levelled against one kind of functions or another very often tend to criticize the methods of approximation, not the choice of functions. For example, least squares fits are almost always oscillatory, for the very good reason that when squared error is minimized, negative and positive errors are taken to be equally acceptable.

Material properties as measured in the laboratory or communicated on paper are normally available as tables of numbers, e.g., values of B tabulated against values of H. The data available may therefore be regarded as a set of sampling points. That is, the data consist of a set V of values of the function $v(x)$ to be modelled, at sampling points x within a point set P,

$$V = \{v(x_j) \mid x_j \text{ in set } P\}, \tag{12}$$

where

$$P = \{x_j \mid j = 1, \ldots, N\}. \tag{13}$$

The construction of an approximate curve $a(x)$ may be regarded as a mapping of the samples V through an interpolation operator $T(P)$, different for every set P of points:

$$a(x) = T(P)\, V. \tag{14}$$

This curve fitting process may be performed according to a variety of criteria. That is to say, representations of the form

$$a(x) = \sum_{i=1}^{N} q_i f_i(x) \tag{1}$$

are sought, such as will provide the best possible approximation to $v(x)$ with respect to some suitable criterion, within the linear manifold A,

$$A = \text{span}(F), \tag{15}$$

where

$$F = \{f_i \mid i = 1, \ldots, N\}. \tag{16}$$

In this view, the essential part of any interpolation process $T(P)$ is the derivation of a set $\{q_i \mid i = 1, \ldots, N\}$ of coefficients, for use in equation (1). The interpolation is then easily completed by substitution of the desired value of x. Many different kinds of $T(P)$ exist, of course, all based on different best-fit criteria.

Curve Modelling

In the following, the theoretical background to the most common approximating processes is sketched out, the specific method used by MagNet serving as an example.

MagNet uses six-piece cubic Hermite interpolation polynomials to model the material reluctivity as a function of flux density squared. That is, every curve built by MagNet is sectionally cubic in the squared flux density B^2. The six sections are taken to be of equal lengths in B^2. Choosing sections of equal length speeds subsequent computations and minimizes data storage space. It uses relatively few sections in the lower (straight) part of the mean magnetization curve, but many sections around and above the knee-point, generally a wise procedure to adopt.

Cubic Hermite interpolation polynomials on six segments define fourteen continuous basis functions with continuous derivatives over the whole range. Seven are interpolative (in the usual Lagrangian sense) on the seven points that define the six intervals in B^2. The remaining seven are interpolative in the first derivative on the same point set. Consequently, each curve is completely specified by giving its value and slope at the seven segment endpoints.

An interpolation operator T maps given value sets V into functions $a(x)$ within the manifold A. One such operator may be constructed on the basis of a least squares approximation process. While various other kinds of approximation are possible, the least squares approach currently enjoys wide popularity. The approximate curve $a(x)$ is always given as a weighted combination of the fourteen basis polynomials f_i,

$$a(x) = T \, V = \sum_{i=1}^{14} h_i f_i (x) , \qquad (17)$$

where the coefficients h_i are determined by minimizing the difference between $a(x)$ and the given values V in a least squares sense:

$$\sum_{i=1}^{14} [a(x_i) - v(x_i)]^2 = 0 . \qquad (18)$$

The requisite minimum is easily found, by solving the matrix equation

$$\sum_{i=1}^{14} \sum_{j=1}^{14} f_i (x_j) \, f_k (x_j) \, h_i = \sum_{j=1}^{14} f_k (x_j) \, v(x_j) . \qquad (19)$$

Its solution determines the interpolation operator T as a mapping from the values v_i to the coefficients h_i. Subsequently, $a(x)$ is easily determined by substitution in (17).

The independent variable x in MagNet is the square of flux density B. Modelling of mean magnetization curves and other material properties is done similarly, except that the mean magnetization curve is not actually modelled and stored by constructing an approximate representation of H, but rather of the reluctivity H/B.

Curve Editing

To edit a curve, it is first discretized using a sampling operator S. The domain of S is the manifold A spanned by the approximating functions, and its range is a set V of point values within the domain of $T(P)$. When a user chooses points on a curve for data entry, he effectively establishes such an operator. When editing of curves is required, MagNet uses a sampling operator S which produces numeric values at 19 sampling points. A mild excess of data (19 points to define 14 degrees of freedom) reduces sensitivity to roundoff error accumulation in any subsequent numerical calculations.

Editing operations may be regarded as a set of transformations $E(p)$ which map sample sets into sample sets, in a manner determined by the sequence p of editing instructions issued by the user. Suppose $a(x)$ is an approximated curve already on file. Editing operations undertaken on it will in general produce a related, but different, curve $b(x)$. The process of editing may be represented by

$$V = S\ a(x), \tag{20}$$

$$V' = E(p)\ V, \tag{21}$$

$$b(x) = T\ V'. \tag{22}$$

In other words, the continuous function $a(x)$ is first given a discrete representation by the sampling operator S. The set V of point values is then altered by the editing process $E(p)$ to form a new set V'; and the edited set of values is finally mapped onto a new continuous curve in A, by the interpolation operator $T(P)$.

The sampling operator S and the interpolation operator T cannot be chosen entirely independently. For the editor to be useful, a large variety of instruction sequences p must be possible; but at the very least, $E(p) = I$ (the identity operation) must be included among the possible editor functions. In that case, no change should result in the original curve. That is, the sampling operator S must be such that $T(P)\ S$ is a projection operator on the manifold A spanned by the approximating functions:

$$T(P)\ S\ a = a. \tag{23}$$

To put the matter simply, a null editing session (generation of sampling points followed by reconstruction of the numerical approximation from the same points) must exactly reproduce the original curve.

Interactive computing systems work best if the user is able to change his mind freely, and to recover from mistakes gracefully. This goal can best be achieved by allowing him full liberty of movement within the universe of possible actions defined by the system commands and by ensuring that to every possible command there corresponds another (or perhaps a combination of others) capable of undoing the action of the first. A second requirement on the set of possible $E(p)$ is therefore the need for inverses to exist. This need is met by ensuring that there exists at least one pair of possible operations which are inverses of each other: creation of a new data point (i.e., enlargement of the set V by one point) and deletion of a point. These two suffice to allow editing, and many sophisticated editor operations may be regarded for theoretical purposes as being simply compositions of numerous deletions and creations. Alteration of a data value, for example, may be regarded as equivalent to deletion, followed by creation of a new data point possessing the desired value.

The Potential Equations of Magnetics

Practically all currently available CAD systems for magnetics and electromagnetics represent physical problems in terms of electric or magnetic potentials. The mathematical physics underlying potential theory, and potential theory as it applies to problems of magnetics, are reviewed in this chapter, with a few brief examples to illustrate the classes of problems to which one formulation or another might apply. The mathematical tools which are subsequently used to solve potential problems are generally numerical. This chapter is intended as a convenient summary review, not as an introduction to electromagnetics.

Electromagnetic Fields and Potentials

Analysis of magnetic fields for design purposes is commonly carried out in terms of two kinds of potential functions: the magnetic vector potential and the magnetic scalar potential. In contrast, the classical description of electromagnetic fields is framed in terms of the Maxwell field equations, a system of partial differential equations involving the fields and flux densities themselves. The potential equations commonly employed in magnetic field analysis and CAD, which are easier to solve than Maxwell's equations, are derived here from the latter. The derivation given here is fairly complete, but is presented in a quite concise form; readers not well acquainted with electromagnetic theory are well advised to consult one of the many good textbooks in that area.

Maxwell's Equations

A complete physical description of any electromagnetic field can be given in terms of five field vectors, two magnetic and three electric. The magnetic vectors comprise the magnetic field **H** and the magnetic flux density **B**, while the three electric vectors are the electric field **E**, the electric flux density **D**, and the current density **J**. There is no magnetic current density.

These quantities are related to the spatial distribution of electric charge density in such a way as to satisfy *Maxwell's equations,*

$$\text{curl } \mathbf{E} = -\frac{\partial}{\partial t}\,\mathbf{B}, \tag{1}$$

$$\text{curl } \mathbf{H} = \mathbf{J} + \frac{\partial}{\partial t}\,\mathbf{D}, \tag{2}$$

$$\text{div } \mathbf{B} = 0, \tag{3}$$

$$\text{div } \mathbf{D} = \rho. \tag{4}$$

These equations represent topological relationships between the fields, in the same sense that the Kirchhoff loop and node equations express the interconnection of branches and loops in a network. There exists a second set of relationships, the *constitutive relations,*

$$\mathbf{B} = \mu\mathbf{H}, \tag{5}$$

$$\mathbf{D} = \epsilon\mathbf{E}, \tag{6}$$

$$\mathbf{J} = g\mathbf{E}. \tag{7}$$

These equations correspond in a certain broad sense to the component descriptions of network analysis; they describe each material medium in the region of interest, in terms of its permeability μ, its permittivity ϵ, and its conductivity g.

The Maxwell equations are obviously partial differential equations, and they cannot be solved unless sufficient boundary conditions are supplied. In the real physical world, there are really no boundary conditions in the mathematical sense, for the universe has no edge. However, there exist conditions which must hold at material interfaces, and which are commonly used to derive boundary conditions adequate for mathematical purposes. In textbooks, the *interface conditions* accompanying Maxwell's equations are usually derived from the equations themselves, so there are exactly as many interface conditions as there are Maxwell equations. At the interface between two regions, say 1 and 2, these are

$$\mathbf{n} \times (\mathbf{E}_1 - \mathbf{E}_2) = 0, \tag{8}$$

$$\mathbf{n} \times (\mathbf{H}_1 - \mathbf{H}_2) = \mathbf{J}_s, \tag{9}$$

$$\mathbf{n} \cdot (\mathbf{D}_1 - \mathbf{D}_2) = \sigma, \tag{10}$$

$$\mathbf{n} \cdot (\mathbf{B}_1 - \mathbf{B}_2) = 0. \tag{11}$$

Here $\mathbf{n}$ is the surface normal vector directed from region 2 into region 1, σ is the surface electric charge density, and $\mathbf{J}_s$ represents surface electric current density.

Like Maxwell's equations themselves, the four interface conditions constitute symmetric pairs in several senses. Two express a continuity requirement for the normal vector components at an interface; the other two deal with tangential components. Two are statements about field vectors; two about flux densities. Two form a pair in electric quantities, the other two in magnetic quantities. Two are homogeneous, implying that there are no individual magnetic poles and no magnetic currents; two are inhomogeneous and may thus be regarded as saying that electric charges and their motion are the basis of the electromagnetic field.

When Maxwell's equations and the constitutive relations for some particular problem are taken together, they constitute a system of seven equations in the five vectors $\mathbf{H}$, $\mathbf{B}$, $\mathbf{E}$, $\mathbf{D}$, $\mathbf{J}$. Substitution of the constitutive relations into the Maxwell equations can reduce the number of vectors to two, one electric and one magnetic. Such a reduction of complexity is essential if attempts at solution are to be successful in practical cases. Indeed, as outlined below, it is usual to strive for much greater complexity reduction still. Complexity can be reduced in three ways: by combining equations so as to eliminate variables, by defining potential functions from which the fields can be derived subsequently, and by seeking simplifying assumptions valid for particular problems. The first two approaches are properly part of mathematical engineering science and will be dealt with in this chapter. The third, seeking simplifying assumptions and approximations, is the basis of the art of engineering. It is by far the most important of the three, and is treated extensively in other chapters of this book.

The Wave Equations

The constitutive relations that describe materials can be combined with the Maxwell equations which relate the fields themselves, so as to yield just two partial differential equations involving only two vector quantities. If the two remaining vector quantities are chosen to be the fields $\mathbf{E}$ and $\mathbf{H}$, it is in fact possible to arrive at two equations which involve only one field vector each. In some problems the two equations can then be solved independently of each other, so that the necessary mathematical work can be reduced to the solution of one partial differential equation in one vector quantity at a time. The differential equation involved in this case is the *wave equation* and is arrived at as follows.

Let the curl operator be applied to both sides of the magnetic curl equation (2) above. In the resulting equation,

$$\operatorname{curl} \operatorname{curl} \mathbf{H} = \operatorname{curl} \mathbf{J} + \operatorname{curl} \frac{\partial \mathbf{D}}{\partial t}, \tag{12}$$

let the vectors $\mathbf{J}$ and $\mathbf{D}$ be eliminated by making use of the material constitutive relations (6) and (7). Taking the material permittivity ϵ and its conductivity g to be constants for simplicity, there results a differential equation of the second order:

$$\text{curl curl } \mathbf{H} = g \text{ curl } \mathbf{E} + \epsilon \frac{\partial}{\partial t} \text{ curl } \mathbf{E}. \tag{13}$$

The electric field $\mathbf{E}$ may next be eliminated, by making use of the Maxwell electric curl equation (1), with the result

$$\text{curl curl } \mathbf{H} = -g \frac{\partial \mathbf{B}}{\partial t} - \epsilon \frac{\partial^2 \mathbf{B}}{\partial t^2}. \tag{14}$$

The remaining constitutive relation may then be employed to recast this equation entirely in terms of the magnetic field $\mathbf{H}$, without explicit reference to the flux density $\mathbf{B}$, as

$$\text{curl curl } \mathbf{H} = -\mu g \frac{\partial \mathbf{H}}{\partial t} - \mu \epsilon \frac{\partial^2 \mathbf{H}}{\partial t^2}. \tag{15}$$

This equation, it should be noted, involves only the field vector $\mathbf{H}$, and can therefore be solved without recourse to the other vectors, provided it is possible to state boundary conditions in terms of $\mathbf{H}$ only.

Equation (15) can be recast in an equivalent form which is no different mathematically, but which is conventionally employed for aesthetic reasons. The recasting involves the common vector identity, valid for any sufficiently differentiable vector $\mathbf{X}$,

$$\text{curl curl } \mathbf{X} = \text{grad div } \mathbf{X} - \nabla^2 \mathbf{X}. \tag{16}$$

Using this identity to eliminate the double curl on the left of (15), and noting that the divergence of $\mathbf{H}$ vanishes (because the divergence of $\mathbf{B}$ always vanishes), there finally results

$$\nabla^2 \mathbf{H} - \mu g \frac{\partial \mathbf{H}}{\partial t} - \mu \epsilon \frac{\partial^2 \mathbf{H}}{\partial t^2} = 0. \tag{17}$$

This equation is commonly called the *wave equation* in $\mathbf{H}$. Its name is somewhat inaccurate; although first used to establish the existence of electromagnetic waves, it in fact encompasses static and diffusive fields as well as propagating waves.

A similar, and strikingly symmetric, development leads to a wave equation in the electric field $\mathbf{E}$. Instead of beginning with the magnetic curl equation, the procedure is to start by taking curls on both sides of the electric curl equation (1),

$$\text{curl curl } \mathbf{E} = -\frac{\partial}{\partial t} \text{ curl } \mathbf{B}, \tag{18}$$

then to eliminate $\mathbf{H}$ by means of the constitutive relation (5),

$$\text{curl curl } \mathbf{E} = -\mu \frac{\partial}{\partial t} \text{ curl } \mathbf{H}, \tag{19}$$

next to appeal to the magnetic curl equation (2) so as to convert entirely to electric quantities,

$$\text{curl curl } \mathbf{E} = -\mu \frac{\partial \mathbf{J}}{\partial t} - \mu \frac{\partial^2 \mathbf{D}}{\partial t^2}, \tag{20}$$

and finally to eliminate $\mathbf{J}$ and $\mathbf{D}$ in favor of $\mathbf{E}$, by means of the two remaining constitutive relations, thus arriving at

$$\text{curl curl } \mathbf{E} = -\mu g \frac{\partial \mathbf{E}}{\partial t} - \mu \epsilon \frac{\partial^2 \mathbf{E}}{\partial t^2}. \tag{21}$$

This equation too may be recast in more conventional notation as

$$\nabla^2 \mathbf{E} + \text{grad div } \mathbf{E} = -\mu g \frac{\partial \mathbf{E}}{\partial t} - \mu \epsilon \frac{\partial^2 \mathbf{E}}{\partial t^2}. \tag{22}$$

However, this time the divergence term in (16) does not vanish, because electric charges may exist anywhere. Hence the wave equation in $\mathbf{E}$ is inhomogeneous,

$$\nabla^2 \mathbf{E} - \mu g \frac{\partial \mathbf{E}}{\partial t} - \mu \epsilon \frac{\partial^2 \mathbf{E}}{\partial t^2} = \text{grad}\left(\frac{\rho}{\epsilon}\right). \tag{23}$$

The pair of wave equations (17)–(22) is equivalent to Maxwell's equations and the material constitutive relations, at least in simple and well behaved media (the derivation above involves exchange of differential operators and material properties on several occasions). No source terms appear anywhere, other than the electric charge density ρ, indicating once again that the electromagnetic field is entirely caused by electric charges.

The wave equations involve only one vector each. Unfortunately, it is not often possible to give boundary conditions in terms of the same vector. Since solution without boundary conditions is not possible, problem complexity is not reduced if the equation is simplified but the boundary conditions are not. An alternative formulation in two variables, the *potentials*, will therefore be considered next.

Electric and Magnetic Potentials

The *magnetic vector potential* and its accompanying *electric scalar potential* are quantities from which the electromagnetic field vectors can be derived, and which have the mathematical advantage that boundary conditions are more often easily framed in the potentials than in the fields themselves. The potentials are therefore a principal working tool in CAD.

The *vector potential* is commonly defined to be a vector $\mathbf{A}$ such that the flux density $\mathbf{B}$ is derivable from it by the curl operation:

$$\text{curl } \mathbf{A} = \mathbf{B}. \tag{24}$$

There is nothing peculiarly electromagnetic in this definition. A similar potential could be associated with any other solenoidal field. The vector identity

$$\text{div curl } \mathbf{A} = 0 \tag{25}$$

is valid for any sufficiently differentiable vector $\mathbf{A}$. Hence curl $\mathbf{A}$ is always nondivergent, and a potential field $\mathbf{A}$ can always be chosen so that its curl equals some prescribed field. But (24) does not define $\mathbf{A}$ uniquely, because a second vector $\mathbf{A}'$ may be defined whose curl is also equal to $\mathbf{B}$. $\mathbf{A}$ and $\mathbf{A}'$ must then be related by the gradient of a scalar function y,

$$\mathbf{A} - \mathbf{A}' = - \text{grad } y, \tag{26}$$

because, for any sufficiently differentiable scalar y, the identity

$$\text{curl grad } y = 0 \tag{27}$$

holds. Applying equation (27) to (26), there results

$$\text{curl } (\mathbf{A} - \mathbf{A}') = 0 \tag{28}$$

showing that the flux density $\mathbf{B}$ derived from either vector potential is the same, even though the potentials are not the same.

The magnetic vector potential needs to be accompanied by an electric scalar potential, in order to eliminate the field vectors entirely from the electromagnetic equations. On combining the Maxwell electric curl equation (1) with the definition (24) of the magnetic vector potential, there is obtained

$$\text{curl } \mathbf{E} = - \text{curl } \frac{\partial \mathbf{A}}{\partial t} \tag{29}$$

or, rewriting slightly,

$$\text{curl } \left[\mathbf{E} + \frac{\partial \mathbf{A}}{\partial t} \right] = 0. \tag{30}$$

Now equation (30) merely states that the vector quantity in parentheses is irrotational. By the general identity (24), it can therefore be represented as the gradient of a scalar V. Hence, the electric field is derivable from the potentials by

$$\mathbf{E} = - \frac{\partial \mathbf{A}}{\partial t} - \text{grad } V. \tag{31}$$

V is called the *electric scalar potential*. In static fields, it reduces to the familiar electric potential of electrostatics, as may be readily verified by removal of the time derivative term.

As observed above, the vector potential **A** is not unique; whole families of different potentials **A** can yield the same electromagnetic field. In a similar way, V is not unique either. Any desired constant could be added to it without effect on the electromagnetic field that results, because the gradient of any constant is always zero.

The Vector Potential Wave Equation

The electric and magnetic potentials would be of no great use if they did not simplify problem solving. As presented above, they appear haphazardly scattered about in the field equations, so it is natural to inquire whether equations can be set up which involve only one potential each, similar to the wave equations (17) and (22) in the magnetic and electric fields. Indeed they can, and it is found that their form is again precisely that of wave equations.

To obtain the wave equations in terms of potentials, let the curl operator be applied to both sides of equation (24), which defines the magnetic vector potential:

$$\text{curl curl } \mathbf{A} = \text{curl } \mathbf{B}. \tag{32}$$

The Maxwell magnetic curl equation (2) may be used to rewrite the right-hand side, so that

$$\text{curl curl } \mathbf{A} = \mu \left[\mathbf{J} + \frac{\partial \mathbf{D}}{\partial t} \right], \tag{33}$$

and the constitutive relations can be applied, so as to substitute the electric field **E** for the current density **J** and the flux density **D** in both right-hand terms:

$$\text{curl curl } \mathbf{A} = \mu \left[g\mathbf{E} + \frac{\partial \mathbf{E}}{\partial t} \right]. \tag{34}$$

But the electric field **E** is related to the potentials **A** and V, and is derivable from them through equation (31). Substituting (31), equation (34) becomes

$$\text{curl curl } \mathbf{A} = -\mu g \frac{\partial \mathbf{A}}{\partial t} - \frac{\partial^2 \mathbf{A}}{\partial t^2} - \mu g \text{ grad } V - \mu \epsilon \text{ grad } \frac{\partial V}{\partial t}. \tag{35}$$

This equation can be recast in the conventional form of a wave equation, making use once again of the vector identity (16). There results

$$-\nabla^2 \mathbf{A} + \mu g \frac{\partial \mathbf{A}}{\partial t} + \mu \epsilon \frac{\partial^2 \mathbf{A}}{\partial t^2} = -\text{grad} \left[\text{div } \mathbf{A} + \mu g V + \mu \epsilon \frac{\partial V}{\partial t} \right]. \tag{36}$$

Now the left-hand terms of this equation involve only the magnetic vector potential $\mathbf{A}$, and have exactly the form expected of a wave equation. However, the right-hand side is rather messy. The mess can be cleared up by exploiting the fact that $\mathbf{A}$ is not unique, and that its divergence may be specified at will. One popular choice is

$$\text{div } \mathbf{A} = -\frac{\partial V}{\partial t}. \tag{37}$$

This equation is known as the *Lorentz condition*. It turns (36) into the rather more elegant inhomogeneous wave equation

$$-\nabla^2 \mathbf{A} + \mu g \frac{\partial \mathbf{A}}{\partial t} + \mu \epsilon \frac{\partial^2 \mathbf{A}}{\partial t^2} = -\mu g \text{ grad } V. \tag{38}$$

This form is directly useful in a variety of physical problems, some of which are sketched out further below.

Potential Problems of Electromagnetics

The vector and scalar potential functions defined in the section above are the principal working tools in most electromagnetics CAD. In the following, they are interpreted physically through illustrative problems. Some of the principal approximations which allow practical cases to be treated are also introduced here, although the main body of such techniques will be treated in other chapters of this book.

Interpretation of the Potentials

The electric and magnetic potentials lend themselves to direct physical interpretation in most cases. Some simple examples of potential problems may well illustrate these interpretations, and may suggest others applicable to more complicated problems.

The electric scalar potential V is exactly the potential familiar from dc circuit theory, where potential differences between circuit nodes generally are called *voltages between nodes*. The terms *voltage* and *potential difference* are not interchangeable, however, because their meanings coincide only in the time-invariant case.

A pair of fine wire loops is shown in Fig. 1. Currents circulate around the left loop. They are driven by the source shown and impeded by the resistance of the wire. The right loop is not connected to the left one, but the magnetic flux of the left loop links the right one. No current flows in the right loop, since it is broken at the terminals. The voltage (the electromotive force) $e(t)$ between the right-hand terminals is given by Faraday's law,

$$e = -\frac{\partial \phi}{\partial t}. \tag{39}$$

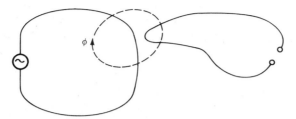

Figure 1. Two fine wire loops linked by magnetic flux.

Now the flux ϕ linking the right-hand loop must be the surface integral of the flux density, taken over the surface S spanned by that loop,

$$\phi = \int_s \mathbf{B} \cdot \mathbf{dS}. \tag{40}$$

But the vector potential **A** must always be such as to satisfy

$$\text{curl } \mathbf{A} = \mathbf{B}, \tag{24}$$

so that (41) may be written

$$\phi = \int_s \text{curl } \mathbf{A} \cdot \mathbf{dS}. \tag{41}$$

According to Stokes' theorem, however, the surface integral of the curl of any vector may be replaced by a line integral of the vector itself, taken around the bounding edge of the surface. The bounding edge of S is clearly the wire that makes up the right-hand loop. Thus the right-hand side of (41) may be converted into a closed line integral expression,

$$\phi = \oint_s \mathbf{A} \cdot \mathbf{ds}, \tag{42}$$

where the line element ds follows the wire direction of the right-hand loop. The general conclusion follows: the line integral of **A** around a closed path always measures the total flux linked by the path.

The electromotive force associated with the right-hand loop in Fig. 1 is given by

$$e(t) = \oint_s \frac{\partial \mathbf{A}}{\partial t} \cdot \mathbf{ds}. \tag{43}$$

In other words, the vector potential **A** measures the flux linking any closed path, and therewith the electromotive force that will be induced in the path if the flux is made to vary with time.

It might be tempting to regard equation (43) as saying that the electromotive force arises from a source distributed around the loop, with every elementary length **ds** of the wire making a contribution of **A · ds** to the total electromotive force. However, a note of caution is due: the integral in (43) spans the whole closed loop. It therefore gives only the total electromotive force due to the entire loop, not of any portion of the loop by itself. This is not to say that the piece-by-piece interpretation is incorrect, merely that equation (43) by itself can prove it neither right nor wrong.

Time-Invariant Potential Equations

If the problem under consideration is static, that is, if all quantities are invariant with time, then all time derivatives in the wave equations vanish. There then remain the two differential equations

$$\nabla^2 \mathbf{A} = -\mu \mathbf{J} \tag{44}$$

and

$$\nabla^2 V = -\rho/\epsilon. \tag{45}$$

These two equations are known as the vector and scalar *Poisson's equations* respectively. They are independent, and they can be solved independently if boundary conditions can be stated in terms of the two potentials separately. This requirement is generally not severe, for in time-invariant problems the magnetic flux causes no electromotive force to appear.

A simple but frequently encountered class of problem is illustrated by Fig. 2. Three conductors embedded in dielectric and placed within a conductive sheath or housing form a cable. A common requirement is to evaluate the cable capacitance per unit length, for which in turn it is necessary to know the spatial distribution of the electric potential V within the dielectric. Since there are no distributed charges in the dielectric material, the potential satisfies a simplified version of (45),

$$\nabla^2 V = 0, \tag{46}$$

known as *Laplace's* equation. The boundary conditions are quite simple in this case. Each conductor may be assumed to be at the same potential throughout, since the voltage between two points on any one conductor is likely to be very small compared to the voltage between two different con-

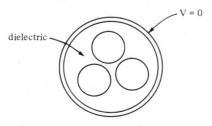

Figure 2. Cross-sectional view of a high-voltage cable. The conductors and the cable sheath are assumed to be equipotential surfaces; the dielectric is taken to be perfectly nonconductive.

ductors. Setting each conductor boundary to be a boundary of fixed potential, as in Fig. 2, the potential V can be solved for, the capacitance and any other derived quantities being computed subsequently from the potential solution.

The problem of Fig. 2 is mathematically two-dimensional, for in a long straight cable it may reasonably be assumed that there are no longitudinal potential variations. At the present state of the art, solution of two-dimensional potential problems of this kind can be carried out routinely. The corresponding type of three-dimensional problem (e.g., a cable termination or sharp bend) causes grave difficulty in result interpretation, computation, and even in problem formulation. Consequently, much of the designer's art resides in decomposing three-dimensional reality into sets of two-dimensional problems, each of which will yield adequate results for some aspect of the problem.

Two-Dimensional Vector Potential Problems

Two-dimensional problems mathematically similar to the cable problem described above are often encountered in the magnetic vector potential. By way of illustration, Fig. 3 shows part of the cross-section of an electric machine rotor. The rotor surface has longitudinal slots cut into it, with current-carrying conductors laid in each slot. The rotor length in its axial direction is much greater than the lateral dimensions of a slot, so this problem too may be regarded as two-dimensional, with the current density $\mathbf{J}$ possessing only a longitudinal (z-directed) component. According to equation (44), the vector potential $\mathbf{A}$ then possesses only a single longitudinal component also, and (44) assumes the simpler form of a scalar Poisson's equation,

$$\nabla^2 A = -\mu J, \tag{47}$$

where A and J now denote the longitudinal components of the vectors $\mathbf{A}$ and $\mathbf{J}$.

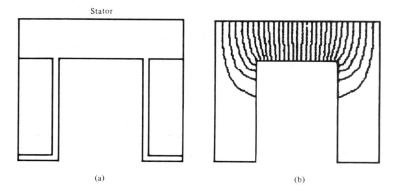

Figure 3. (a) An electric machine rotor tooth, flanked by slots enclosing current-carrying conductors. (b) Flux distribution around the slot and tooth.

To solve for the magnetic vector potential within and near the rotor slot, boundary conditions must be supplied in addition to equation (47). A basis for obtaining suitable boundary conditions can be created applying equation (42) to a somewhat special choice of closed contour. Let a rectangular contour R of integration be chosen as follows: from point P: $(x_P, y_P, 0)$ to the point Q: $(x_Q, y_Q, 0)$, then parallel to the z (longitudinal) axis a unit distance to Q': $(x_Q, y_Q, 1)$, then in a straight line parallel to the $x-y$ plane to P': $(x_P, y_P, 1)$, and finally back to point P in a straight line parallel (but oppositely directed) to the z axis. The contour integral may clearly be written as the sum of four line integrals,

$$\oint_R \mathbf{A} \cdot \mathbf{ds} = \int_P^Q \mathbf{A} \cdot \mathbf{ds} + \int_Q^{Q'} \mathbf{A} \cdot \mathbf{ds} + \int_{Q'}^{P'} \mathbf{A} \cdot \mathbf{ds} + \int_{P'}^P \mathbf{A} \cdot \mathbf{ds}. \quad (48)$$

Because the problem geometry is uniform in the z direction, the vector potential $\mathbf{A}$ is everywhere parallel to the z axis. Thus the contour segments PQ and Q'P' are everywhere at right angles to $\mathbf{A}$, and the integrals over those two segments vanish because their integrands vanish. Along the segment QQ' the vector potential has the value $\mathbf{A}(Q)$ everywhere and is directed exactly along the line segment $\mathbf{ds}$, for $\mathbf{ds}$ is there parallel to the z axis. The integral along that contour segment is then simply $\mathbf{A}(Q)$, the length of segment QQ' being unity. Along the segment $P'P$ a similar argument applies, except that the line segment $\mathbf{ds}$ is now directed in opposition to the z axis; the integral therefore evaluates to $-\mathbf{A}(P)$. For the special contour R, equation (48) thus becomes

$$\oint_R \mathbf{A} \cdot \mathbf{ds} = \mathbf{A}(Q) - \mathbf{A}(P). \quad (49)$$

In accordance with equation (24), all flux lines must lie parallel to the $x-y$ plane in this problem. Suppose now that the contour R is chosen so that

$$\mathbf{A}(Q) = \mathbf{A}(P). \quad (50)$$

The integral in (49) must vanish, implying that the contour R in this case links no flux at all. But that is only possible if points P and Q lie on the same flux line! A flux line may therefore be traced by following equal values of $\mathbf{A}$; and conversely, *a contour of equal values of A is a flux line.*

The slot-conductor problem of Fig. 3 is symmetric about the centerline of each slot, and also symmetric about the centerline of each tooth separating two adjacent slots. Because flux lines encircle conductors, the tooth centerline (which separates two successive slots) must be a flux line, along which the vector potential $\mathbf{A}$ has a fixed value. For convenience, this value may be taken as zero, so that

$$A = 0 \quad \text{along the tooth centerline.} \quad (51)$$

In contrast, the slot centerline must be a line of even symmetry in A, all flux lines must cross it at right angles. Hence the appropriate boundary condition is

$$\frac{\partial A}{\partial n} = 0 \quad \text{along the slot centerline.} \tag{52}$$

Finally, if the surrounding rotor and stator iron are assumed to be infinitely permeable, flux lines must impinge on the iron surfaces at right angles also. The boundary condition applicable there is consequently

$$\frac{\partial A}{\partial n} = 0 \quad \text{along every iron—air boundary.} \tag{53}$$

Boundary conditions in which the potential is prescribed are often referred to as *Dirichlet conditions*, while those in which the normal derivative of potential is prescribed are given the name of *Neumann boundary conditions*. Conditions in which the prescribed values are zero, as in equations (51)–(53), are usually termed homogeneous.

Potentials in Solid Conductors

A similar interpretation to that given above for slim wires holds for potentials in solid conductors. Consider, for example, two parallel current-carrying conductors of square cross-sectional shape, as shown in Fig. 4. The conductors are assumed to be very long, and to terminate in a short circuit formed by a sheet of perfect conductor. A time-varying source is connected between the near ends of the conductors and maintains a prescribed voltage between them.

If the source voltage were to vary with time very slowly indeed, both time derivatives in the wave equation (38) would be absent. In accordance with equation (31), the current in the conductors would then be the result of an essentially static electric field, as expressed by $-grad\ V$. As a result

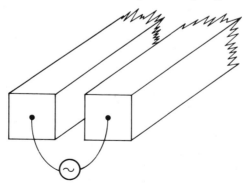

Figure 4. The classical skin effect problem: the conductors carry time-varying, essentially longitudinal, currents.

of Ohm's law, equation (7) above, the current density would be distributed in the conductor as prescribed by

$$\mathbf{J} = -g \ \text{grad} \ V.$$ (54)

In other words, the current density at very low frequencies is distributed in the conductor in very nearly the same fashion as direct currents. The resulting magnetic field must then be such that the vector potential $\mathbf{A}$ satisfies

$$\nabla^2 \mathbf{A} = -\mathbf{J},$$ (55)

since all time derivatives are negligible in this case.

To calculate the skin effect in this conductor pair, time variations must be taken into account, for the physical basis of skin effect is precisely that time variations in the magnetic field (hence in $\mathbf{A}$) give rise to a *back emf* which causes eddy currents in the conductor. These circulating currents augment the source current density in some parts of the conductors, oppose them in others. Skin effect calculations are generally of interest only in the frequency range in which conductive effects far outweigh capacitive ones, i.e., the electric displacement current is much smaller than the conduction current $\mathbf{J}$ within the conductor material,

$$\| \ \mathbf{J} \ \| \ \gg \ \| \ \frac{\partial \mathbf{D}}{\partial t} \ \| \ .$$ (56)

Neglecting the displacement current has the effect of suppressing the second time derivatives in all wave equations, as may be verified by retracing the above developments. The general wave equation in this case reduces to

$$\nabla^2 \mathbf{A} + \mu g \ \frac{\partial \mathbf{A}}{\partial t} = \mu \mathbf{J}.$$ (57)

The magnetic field in the conductor (as determined by the leftmost term) must always be attributable to two electromagnetic force components: one arising directly from the source (the right-hand side) and one due to the self-induced back emf (the time derivative term on the left).

Magnetic Scalar Potentials

In addition to the magnetic vector potential and the electric scalar potential, which arise as a natural pair from the Maxwell equations, several other potential functions are commonly used in magnetics. The best known of these are probably the *magnetic scalar* potentials, which are available in several commercial CAD systems, and which are therefore briefly treated below.

The Total Scalar Potential

Magnetostatics problems are at times viewed as entirely analogous to electrostatics problems, with magnetic flux density corresponding to electric flux density, magnetic field corresponding to electric field, and permeability corresponding to permittivity. In analogy to the electric scalar potential, another potential function, the *magnetic scalar potential*, is then required. This potential function is useful only for problems not involving conduction current density **J**. It is based on the observation that if time variations of electric flux **D** are negligible and there is no conduction current density, **J** = 0, then

$$\text{curl } \mathbf{H} = 0. \tag{58}$$

In this relatively restricted class of cases, **H** may be derived from a potential function,

$$\mathbf{H} = -\text{ grad } \Omega. \tag{59}$$

Multiplying both sides by the material permeability, then taking divergences,

$$\text{div } \mathbf{B} = -\text{ div } \mu \text{ grad } \Omega \tag{60}$$

results. But in accordance with Maxwell's magnetic divergence equation (3), the left-hand side of (60) always vanishes. Hence

$$\text{div } \mu \text{ grad } \Omega = 0 \tag{61}$$

describes magnetostatic fields in current-free regions. The potential Ω is usually referred to as the *magnetic scalar potential*, or the *total magnetic scalar potential*, if it is necessary to distinguish it from the alternative representation to be discussed below.

   The applications in which the magnetic scalar potential has proved useful in the past have most often involved partial investigations of large problems, for example in the determination of how flux lines avoid a bolthole in an iron member. However, the magnetic scalar potential is obviously a true scalar quantity, possessing only a single numerical value at any given space point. In three-dimensional problems in particular, it is therefore more economic to calculate than the vector potential, and results are easier to interpret.

The Carter Coefficient Problem

The magnetic scalar potential is often used for problems involving segments of complicated structures. A typical example is the calculation of the classical Carter coefficient for the air gap of an electric machine. Electric machine rotor surfaces are generally made slotted, with the current-carrying conductors fitted into the slots. If local details of flux distribution around the slots are not of interest, the slotted structure may be considered to behave globally much like a smooth one, but with an air gap of increased size. The Carter coefficient problem asks, in essence, *how large is the smooth air gap globally equivalent to the real gap*, which includes both teeth and slots.

As in the current-carrying slot conductor problem, the structure of Fig. 5 has sufficient symmetry to permit solving over only a limited region. Once again, the problem is two-dimensional. If the problem is formulated in terms of the magnetic scalar potential, equation (61) must be satisfied. In addition, boundary conditions must again be supplied to make the problem solvable. Boundary conditions may be formulated using an approach similar to the vector potential case above, even though the results will of course be different.

The definition of magnetic scalar potential, equation (59), may be used to find the scalar potential difference between any two space points P and Q. Integrating both sides of (59) along some path from P to Q,

$$\int_P^Q \mathbf{H} \cdot \mathbf{ds} = \Omega(Q) - \Omega(P) \tag{62}$$

which says that the difference in magnetic scalar potential corresponds exactly to the *magnetomotive force* between two points as encountered in classical magnetic circuit theory. In static problems, the magnetic scalar potential is thus the dual of the electric scalar potential of circuit analysis.

Suppose now that P and Q are two arbitrary points in a surface of constant magnetic scalar potential. By (62), the magnetic field vector $\mathbf{H}$ must have zero projection on this surface, or, in other words, the scalar equipotential surface must be orthogonal everywhere to lines of $\mathbf{H}$. In isotropic materials, where the magnetic field and flux vectors are parallel, the equipotential surfaces are thus orthogonal to magnetic flux lines.

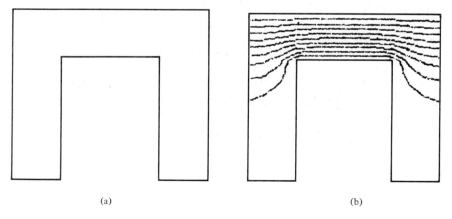

(a)   (b)

Figure 5. (a) Rotor tooth and smooth stator surface of an electric machine, with rotor conductors removed. (b) The scalar equipotentials around the tooth.

If surfaces of constant magnetic scalar potential are always orthogonal to flux lines, and the centerline of a rotor tooth is a flux line, it must also be a line of even symmetry of scalar potential. So must the centerline of a rotor slot, by the same argument. Hence

$$\frac{\partial \Omega}{\partial n} = 0 \quad \text{along tooth and slot centerlines.} \tag{63}$$

On the other hand, if the rotor and stator iron are assumed to be infinitely permeable, then flux lines must always impinge on the iron surfaces at right angles. The iron surfaces must therefore be surfaces of constant magnetic scalar potential:

$$\Omega = \Omega_r \quad \text{on the rotor surface,} \tag{64}$$

and

$$\Omega = \Omega_s \quad \text{on the stator surface.} \tag{65}$$

The nonlinear Laplace equation (61), together with the boundary conditions (62)–(65), forms a well-posed problem which may be solved to determine the scalar potential in the slots and in the machine air gap. From the resulting potential distribution, the Carter coefficient can then be determined.

Reduced Scalar Potentials

The restriction that all problem regions must be source-free is quite severe, since most useful problems in magnetics involve either true current densities or their equivalents in permanent magnets. The notion of a scalar potential can be made more generally useful by removing the requirement that the magnetic field must be irrotational everywhere. Let a general magnetostatic situation be considered, in which both current-carrying conductors and soft magnetic materials are placed arbitrarily in space. The magnetic field in and around the magnetic materials cannot be directly expressed in terms of the scalar potential, because the field is not irrotational in the conductors. An auxiliary problem is therefore defined, by assuming that all magnetic materials have been removed from the problem. Although this auxiliary problem cannot be formulated in terms of the scalar potential either—there are still currents in the conductors—it is an easier problem because there are no magnetic materials anywhere. In many practical cases, such an auxiliary problem can in fact be framed to have an analytic solution. Let the magnetic field of this auxiliary solution be denoted by $\mathbf{H}_a$. Since the sources are the same in both problems,

$$\text{curl } \mathbf{H}_a = \mathbf{J}. \tag{66}$$

But $\mathbf{H} - \mathbf{H}_a$, the difference between the magnetic fields in the real problem and the auxiliary problem, is an irrotational vector everywhere:

$$\text{curl } (\mathbf{H} - \mathbf{H}_a) = 0. \tag{67}$$

Clearly, this field can be given a representation in terms of a scalar potential,

$$\mathbf{H} - \mathbf{H}_a = - \text{ grad } \Omega_r. \tag{68}$$

Because the field to be represented has been reduced by the auxiliary field $\mathbf{H}_a$, the corresponding scalar potential is known as the *reduced scalar potential*. Rewriting and multiplying by the local value of permeability, (68) becomes

$$- \mu \text{ grad } \Omega_r = \mu \mathbf{H}_a - \mu \mathbf{H}. \tag{69}$$

The rightmost term clearly represents the flux density $\mathbf{B}$ in the actual physical problem at hand. Its divergence therefore vanishes, and

$$- \text{ div } \mu \text{ grad } \Omega_r = \text{ div } (\mu \mathbf{H}_a). \tag{70}$$

This nonlinear Poisson equation may now be solved to find the reduced scalar potential. In other words, the solution of a field problem by means of the scalar potentials now consists of three steps: (1) defining and finding the auxiliary field $\mathbf{H}_a$, (2) solving the nonlinear potential equation (70), and then (3) combining the two solutions to find the complete field,

$$\mathbf{H} = \mathbf{H}_a - \text{ grad } \Omega_r. \tag{71}$$

It might be noted in passing that the auxiliary field $\mathbf{H}_a$ could be defined in a large variety of ways, provided only that equation (66) holds. This observation leads to a large number of possible potential formulations, differing in the manner in which $\mathbf{H}_a$ is defined. Several of these have been published, but only the normal reduced scalar potential formulation, as sketched out above, appears to be currently available in standard software packages.

Numerical Instabilities

Scalar potential formulations of magnetic problems at times lead to difficulties with numerical stability of the solutions. The reason is not far to seek. Consider again the general problem of current-carrying conductors and iron parts, arbitrarily disposed in space, and let an additional iron part be introduced at some space point. If the magnetic material is highly permeable, the field $\mathbf{H}$ inside the material will be considerably smaller than its value in the absence of the iron; that is,

$H \ll H_a$. That, however, amounts to saying that the auxiliary field $\mathbf{H}_a$ is of the same order of magnitude as the field derived from the reduced scalar potential. Using the three-step calculation procedure, as given above, there is a considerable risk of error, for although both the auxiliary field and the reduced scalar potential may have been computed with high accuracy, their difference will be small and therefore subject to substantial error. In the air, where the two component fields of equation (71) are widely different, the results may be entirely satisfactory. But in the iron, cancellation of almost equal quantities could render results almost meaningless.

One possible cure to this problem, employed in at least one CAD package, is to use two scalar potentials: the total potential in iron parts, where there is no current, and the reduced potential formulation in air and conductors. At an interface between an iron region T, where the total scalar potential is employed, and an air region R, where the reduced potential is used, the normal component of flux density must be continuous. If $\mathbf{1}_n$ denotes the unit normal vector at the region interface, then

$$\mathbf{1}_n \cdot (\mu_R \text{ grad } \Omega_r - \mu_T \text{ grad } \Omega) = \mathbf{1}_n \cdot \mu_R \, \mathbf{H}_a . \tag{72}$$

Similarly, the tangential components of the magnetic field must be continuous at the region interface:

$$\text{grad } (\Omega_r - \Omega) \times \mathbf{1}_n = \mathbf{H}_a \times \mathbf{1}_n . \tag{73}$$

Clearly, the difficulty here is in constructing software to impose these interface conditions without burdening the software user excessively. Alternative formulations of (72) and (73) in integral form can achieve this task reasonably well; these are used, for example, in the TOSCA package.

Time-Varying Potential Problems

Although some useful problems of magnetics can be stated in time-invariant form, numerous others require solution in the time domain. Complications can then arise in formulating the potential problem, as well as in subsequent computation. The principal mathematical tools used to set up such problems are briefly reviewed in this section.

Choice of Gauge in Diffusion Problems

In many potential problems, including the examples given above, the divergence of the vector potential $\mathbf{A}$ is taken to be specified by

$$\text{div } \mathbf{A} = - \mu \epsilon \frac{\partial V}{\partial t} . \tag{37}$$

For historical reasons, this choice is generally referred to as the *Lorentz gauge* or the *Lorentz condition*. Another choice frequently employed in problem solving is

$$\text{div } \mathbf{A} = 0 \qquad (74)$$

known as the *Coulomb gauge*. It is superficially simple, but it can lead to relatively complicated forms of the wave equation. Using (74), the general wave equation

$$\nabla^2 \mathbf{A} + \mu g \frac{\partial \mathbf{A}}{\partial t} + \mu \epsilon \frac{\partial^2 \mathbf{A}}{\partial t^2} = - \text{grad} \left[\text{div } \mathbf{A} + \mu g V + \mu \epsilon \frac{\partial V}{\partial t} \right], \quad (36)$$

assumes the form

$$\nabla^2 \mathbf{A} + \mu g \frac{\partial \mathbf{A}}{\partial t} + \mu \epsilon \frac{\partial^2 \mathbf{A}}{\partial t^2} = - \text{grad} \left[\mu g V + \mu \epsilon \frac{\partial V}{\partial t} \right], \quad (75)$$

which is definite and clearcut, but complicated by the appearance of a time derivative term in V on the right-hand side. In static problems, the Coulomb and Lorentz gauges do of course coincide, but in time-varying situations they do not. Usually, the Lorentz gauge is then the preferred choice.

Another choice of gauge, useful in some eddy current problems, removes the bracketed quantity in equation (36) entirely by choosing the divergence of vector potential to be given by

$$\text{div } \mathbf{A} = - \mu g V - \mu \epsilon \frac{\partial V}{\partial t}. \qquad (76)$$

With this choice, the wave equation in vector potential becomes homogeneous,

$$- \nabla^2 \mathbf{A} + \mu g \frac{\partial \mathbf{A}}{\partial t} + \mu \epsilon \frac{\partial^2 \mathbf{A}}{\partial t^2} = 0. \qquad (77)$$

It should be noted that the magnetic flux is computed by line integration of $\mathbf{A}$,

$$\phi = \oint \mathbf{A} \cdot \mathbf{ds} \qquad (42)$$

regardless of the choice of gauge. The argument on which equation (40) is based relies on the definition of $\mathbf{A}$ through its curl,

$$\text{curl } \mathbf{A} = \mathbf{B}, \qquad (24)$$

without reference to the divergence of $\mathbf{A}$. Equation (42) therefore remains valid no matter what gauge is used. Correspondingly, lines of constant **A** in two-dimensional problems remain flux lines under all gauges.

The Scalar Wave Equation

It is interesting to note that the electric scalar potential also obeys a wave equation. The Maxwell electric divergence equation,

$$\text{div } \mathbf{D} = \rho, \tag{4}$$

may be restated, using the constitutive relation (6), as

$$\text{div } \mathbf{E} = \rho/\epsilon. \tag{78}$$

However, the electric field is always given by

$$\mathbf{E} = -\frac{\partial \mathbf{A}}{\partial t} - \text{grad } V \tag{31}$$

Substituting (31) into (78), there results

$$\nabla^2 V + \text{div}\frac{\partial \mathbf{A}}{\partial t} = -\rho/\epsilon, \tag{79}$$

which will clearly take on a somewhat different appearance for every different choice of gauge. In fact, the word *gauge* refers to the fact that the pair of potentials (**A**, V) can be "calibrated" or "gauged" in different ways; terms removed from one by altering the gauge reappear in the other in some different form. Under the Coulomb gauge (74), equation (79) clearly assumes the form of the familiar scalar Poisson equation,

$$\nabla^2 V = -\rho/\epsilon, \tag{80}$$

compensating for the rather complicated appearance of the corresponding vector potential wave equation (75). Choosing the Lorentz gauge (37) leads to

$$\nabla^2 V + \mu\epsilon\frac{\partial^2 V}{\partial t^2} = \frac{\rho}{\epsilon}, \tag{81}$$

an undamped wave equation. Finally, taking the gauge to be defined by (76) produces

$$-\nabla^2 V + \mu g\frac{\partial V}{\partial t} + \mu\epsilon\frac{\partial^2 V}{\partial t^2} = \frac{\rho}{\epsilon}, \tag{82}$$

a wave equation symmetric in form with the wave equation that applies to the vector potential,

$$-\nabla^2 \mathbf{A} + \mu g\frac{\partial \mathbf{A}}{\partial t} + \mu\epsilon\frac{\partial^2 \mathbf{A}}{\partial t^2} = 0. \tag{77}$$

Aesthetically attractive because of their symmetry, this pair of potential equations also emphasizes that the electromagnetic field arises entirely from electric charges: the electric equation (82) contains the charge density in its right-hand member, while the corresponding magnetic equation (77) is homogeneous.

Separation of Time and Space Variables

Solution of time-varying problems is in principle more complicated than that of corresponding static cases, because a new solution must be computed at each of a succession of time instants. In linear problems, a repeated sequence of solutions can be avoided by separating the time from the spatial variables prior to numerical solution, then recombining afterward. Unfortunately, this process cannot be carried out in nonlinear cases, so that its usefulness is often limited.

Consider again the homogeneous vector potential equation (77) given above. Suppose it is possible to write its solution in the product form

$$\mathbf{A}(x,y,z,t) = \mathbf{S}(x,y,z)\, T(t),\qquad\qquad(83)$$

where $\mathbf{S}$ is a vector-valued space function not dependent on time, and T is a scalar time function uniform in space. If that is possible, then equation (77) assumes the form

$$T\,\nabla^2\,\mathbf{S} - \mu g\,\mathbf{S}\,T' - \mu\epsilon\mathbf{S}\,T'' = 0,\qquad\qquad(84)$$

where the primes denote differentiation. Since T is in general not zero, (84) may be written as

$$\nabla^2\,\mathbf{S} - \left[\frac{\mu g\,T' + \mu\epsilon\,T''}{T}\right]\mathbf{S} = 0.\qquad\qquad(85)$$

But in this form, only one time-dependent term appears: the bracketed coefficient of $\mathbf{S}$. Since $\mathbf{S}$ is time-independent, equation (85) can be an equation only if the apparently time-dependent bracketed term is not in fact time-dependent at all, but a constant; otherwise, (85) could only hold at selected instants of time. It is therefore proper to set

$$\frac{\mu g\,T' + \mu\epsilon\,T''}{T} = -k^2.\qquad\qquad(86)$$

With this value substituted, (85) becomes

$$(\nabla^2 + k^2)\,\mathbf{S} = 0,\qquad\qquad(87)$$

which is a differential equation in space only. Correspondingly, (86) may be written as

$$\mu\epsilon\,T'' + \mu g\,T' + k^2\,T = 0,\qquad\qquad(88)$$

a differential equation in time only. The former is usually called the *vector Helmholtz equation*, because the differential operator which precedes **S** in (87) is called the *Helmholtz operator*. The latter, equation (88), is a familiar enough form: an ordinary differential equation of second order with constant coefficients.

The pair of independent equations (87)–(88) can be solved separately, provided boundary and initial conditions are such as to keep time and space separated. In other words, the spatial boundaries of the problem region must be describable by equations in the form

$$h_m(x,y,z) = 0 \qquad m = 1, \ldots, N \tag{89}$$

in order for the Helmholtz equation to be solvable separately from the time differential equation (88). But if the boundaries could only be described by equations of the form

$$h_m(x,y,z,t) = 0 \qquad m = 1, \ldots, N \tag{90}$$

then the time and space problems would be interdependent, even though the differential equations (87)–(88) appear independent. The reason is simple: what must be solved is not the differential equation, but rather the boundary value problem which comprises both the differential equation and its associated boundary conditions. Both must allow separation of time and space in order to allow solution of the two problems separately. Fortunately, cases in which separation is possible are quite commonly encountered.

If the boundary of some object or region is describable by an equation like (90), boundary points *(x,y,z)* must move with time *t*. In other words, (90) describes boundaries in motion. Thus a general conclusion may be drawn: *separation of variables works for linear problems wherever all problem boundaries remain static.* It cannot be applied to solve problems involving moving objects, nor to problems in which the material properties are time-varying or nonlinear.

Sinusoidal Time Variations

A particularly interesting special case of the variable separation technique occurs when the time variation is sinusoidal and takes place in a steady state. In that case, it is often convenient to use complex variables. The vector potential **A** is then assumed to have a product form similar to that given by (83), but this time it is understood that the time function is a complex exponential:

$$\mathbf{A}(x,y,z,t) = \mathrm{Re}\,[\mathbf{S}(x,y,z)\,\exp(j\omega t)], \tag{91}$$

while **S***(x,y,z)* is a complex space function. Keeping with the practice conventional in circuit theory, both the real part indication Re[ ] and the time exponential factor are suppressed; it is understood that they are to be reintroduced prior to actually calculating any true time functions.

This time, the Helmholtz equation which results will have a complex solution, and the constant $-k^2$ will in general be complex also. In mathematical principle, the procedure of solving the problem appears almost indistinguishable from the real case. In practice, substantial differences arise. The computational effort required to solve a complex set of equations is of course greater than that for a similar-sized real one. However, this aspect of the problem is relatively insignificant. Serious problems are encountered in postprocessing because a real quantity can be represented in graphs and curves fairly directly, but a complex quantity cannot. Such difficulties are considered in detail in the chapters on solution postprocessing; so far as the potential-theoretic questions are concerned, the matter is very straightforward indeed.

Rotational and Translational Symmetries

While the physical world is three-dimensional so that engineering analysis is always ultimately concerned with three-dimensional objects, the realities of both representation and computation make it attractive to seek two-dimensional simplifications. Two distinct approaches are possible. First, the problem can be modified to fit within a particular two-dimensional framework. Secondly, it may be possible to select a coordinate system in which one space variable can be split off from the other two by a process of separation of variables.

In the foregoing, the potential equations relevant to magnetics have been developed and presented in an essentially coordinate-independent three-dimensional notation. In this section, the same equations will be examined briefly as they appear in some conventional coordinate systems, with particular emphasis on the two- dimensional forms that result from rotational or translational symmetries in the boundary shapes.

Translationally Uniform Problems

Many practical engineering problems involve geometric shapes invariant in one direction. Simple examples already mentioned above include transmission lines of all kinds, and electric machine conductors in slots. Almost anything produced by milling machines or extrusion techniques is likely to possess a shape uniform in the direction parallel to the fabrication process, and hence might fit into the framework discussed here.

Let z denote the Cartesian coordinate direction in which the structure to be considered is invariant in size and shape. To put the matter more mathematically, let all the boundaries of the problem considered be such as to be describable by

$$h_m(x,y) = 0 \qquad m = 1, \ldots, N. \tag{92}$$

Now the Laplacian operator in Cartesian coordinates may be written formally as the combination of a two-dimensional Laplacian operator with a second derivative in the z direction,

$$\nabla_3^2 = \frac{\partial^2}{\partial x^2} + \frac{\partial^2}{\partial y^2} + \frac{\partial^2}{\partial z^2},$$

$$\nabla_2^2 = \frac{\partial^2}{\partial x^2} + \frac{\partial^2}{\partial y^2}. \tag{93}$$

The subscripts 2 and 3 are here attached to the Laplacian operators in order to distinguish two- and three-dimensional cases. In general, this notation will be used wherever there is any danger of confusion. Where no ambiguity can arise, the subscripts will be omitted.

The similarity of equation (92) to (89) suggests that a variable separation process might be appropriate. Thus, set

$$\mathbf{S}(x,y,z) = \mathbf{P}(x,y)\, Z(z), \tag{94}$$

where $\mathbf{P}$ is a vector quantity, Z is a scalar. Let (94) be substituted into the Helmholtz equation (87). With the Laplacian written in separated form, as in (93), there immediately results

$$Z\, \nabla^2 \mathbf{P} + \mathbf{P}\, Z'' + k^2\, \mathbf{P}Z = 0. \tag{95}$$

Since the longitudinal function $Z(z)$ cannot in general be zero (or else the problem has the solution $\mathbf{S} = 0$ everywhere!), it is permissible to divide by Z, thus obtaining

$$\nabla^2 \mathbf{P} + (Z''/Z)\, \mathbf{P} + k^2\, \mathbf{P} = 0. \tag{96}$$

Separation is now possible, following exactly the same argument as applied to (85) above. Since only the middle term on the left-hand side of equation (96) involves z, the factor multiplying $\mathbf{P}$ must in fact be a constant, say $-m^2$. Thus there result two equations: an ordinary differential equation in z only,

$$Z'' + m^2\, Z = 0, \tag{97}$$

and another, still a partial differential equation but involving only the remaining two space coordinates,

$$(\nabla^2 + k^2 - m^2)\, \mathbf{P} = 0. \tag{98}$$

The latter equation of course requires boundary conditions which do not involve z. But that is exactly what equation (92) promises!

The ordinary differential equation (97) has the usual possible solutions to be expected of a second-order equation with constant coefficients.

These may be of three distinct kinds, as is easily seen by substitution of the trial function $Z = \exp(pz)$ into (97). There immediately results

$$(m^2 + p^2)\, \exp(pz) = 0, \tag{99}$$

and it follows that (97) will be satisfied if and only if

$$m^2 = -p^2. \tag{100}$$

Three interesting cases arise. First, if m^2 is a positive real number, p is purely imaginary; Z is therefore a combination of sine and cosine terms. The solution then consists of a Fourier series in z, in which each term is multiplied by a distinct solution of (98) as its coefficient,

$$\mathbf{S} = [\mathbf{P}_{ci}\,(x,y)\, \cos\, iz + \mathbf{P}_{si}\,(x,y)\, \sin\, iz]. \tag{101}$$

Secondly, there exists the possibility that $m = p = 0$. The solution S is then *translationally invariant*. The third possibility is that m^2 is a negative number, i.e., that m is imaginary. It is evident from (100) that in this case p must be real, but may be either positive or negative. Z is therefore a combination of exponential terms with positive and negative exponents, more conveniently expressed in terms of hyperbolic functions:

$$\mathbf{S} = \sum_i [\mathbf{P}_{Ci}\,(x,y)\, \cosh\, iz + \mathbf{P}_{Si}\,(x,y)\, \sinh\, iz]. \tag{102}$$

This latter case arises less frequently than the first two; but all three do occur in applications. The most common case by far is that of solutions uniform in z, that is, $m = 0$. The slot-conductor problem considered above is of precisely this nature.

It should be noted that the above separation can always be effected, provided the boundary conditions in the $x-y$ plane can be stated in a form independent of z, as in equation (100). There is no requirement that the right-hand side (the source term) be independent of z. In other words, *separation is feasible if material boundaries are translationally invariant*, even if excitations are not.

In Cartesian coordinates, the above argument holds equally well for vector as well as scalar potentials, because the Laplacian operators applicable to vectors and scalars are formally identical in Cartesians. The same is not true in cylindrical coordinates, hence in problems with axial symmetry. Those very common cases therefore merit separate discussion, as below.

Rotationally Uniform Scalar Problems

Practical engineering problems very frequently involve either translational or rotational uniformity of region shape—perhaps because a great many devices are fabricated by using milling machines or lathes, which produce longitudinally or rotationally uniform cuts! Analysis correspondingly proceeds by starting with either Cartesian or cylindrical coordinates. In cylindrical coordinates, the homogeneous Helmholtz equation reads

$$\frac{1}{r} \frac{\partial}{\partial r} \frac{\partial u}{\partial r} + \frac{1}{r^2} \frac{\partial^2 u}{\partial \phi^2} + \frac{\partial^2 u}{\partial z^2} + k^2 u = 0, \qquad (103)$$

where u is the scalar variable desired. Clearly, this equation is not formally identical to its Cartesian counterpart.

The variable u of equation (103) is not necessarily uniform in the azimuthal direction. If it does vary, it can be treated by separation of variables, in a manner similar to the longitudinal variable above. The separation proceeds once again by substituting

$$u(r,\phi,z) = U(r,z)\, \Phi(\phi) \qquad (104)$$

into (103). There is obtained, by a process similar to (95)–(97), the pair of equations

$$\Phi'' + m^2\, \Phi = 0 \qquad (105)$$

and

$$\frac{1}{r} \frac{\partial}{\partial r} \left[r \frac{\partial U}{\partial r} \right] + k^2 U - \frac{m^2}{r} U + \frac{\partial^2 U}{\partial z^2} = 0. \qquad (106)$$

The first of these, equation (105), is formally identical to (96) and will have formally identical solutions. The only practical difference is that, in the axisymmetric case, solutions must repeat themselves after every full rotation, and aperiodic solutions (hyperbolic functions) will rarely if ever arise. The other equation (106) is sometimes called a *generalized Bessel equation* because it bears a close resemblance to the classical Bessel equation

$$\frac{1}{r} \frac{d}{dr} \left[r \frac{dU}{dr} \right] + k^2 U - \frac{m^2}{r} U = 0, \qquad (107)$$

with the difference that (106) is a partial differential equation containing a z-directed derivative. Because the Bessel equation only differs from (106) in the longitudinal direction, but contains identical radial derivatives, it might be expected that the radial behavior of its solutions near $r = 0$ will resemble the behavior of Bessel functions in that neighborhood, and this expectation is indeed borne out in practice. The dominant influence near the axis is exerted by the term $1/r$, containing m^2. For different values of m, the solutions therefore turn out to have quite different characteristics. If $m = 0$, the singular term is absent, so that solutions are bounded and smooth; for $m = 0$, however, they are unbounded at $r = 0$.

An important special case is that of true axial symmetry, in which the boundary conditions, sources (if any), and solutions are all invariant with the angle ϕ. In this case, $m = 0$, and the general equation (106) reduces to

$$\frac{\partial^2 U}{\partial r^2} + \frac{1}{r} \frac{\partial U}{\partial r} + k^2 U + \frac{\partial^2 U}{\partial z^2} = 0, \qquad (108)$$

whose solutions are well behaved near $r = 0$. Smooth and bounded behavior is to be expected on physical grounds, for U could well represent a scalar potential and therefore should have a unique, well defined, value at any point on the axis.

Rotationally uniform problems of course need not be source-free. With sources included, the Helmholtz equation of (103) becomes

$$\frac{1}{r} \frac{\partial}{\partial r} \left[r \frac{\partial u}{\partial r} \right] + \frac{1}{r^2} \frac{\partial^2 u}{\partial \phi^2} + \frac{\partial^2 u}{\partial z^2} + k^2 u = s, \qquad (109)$$

with $s(r, \phi, z)$ the prescribed source density. To solve the inhomogeneous equation, it is best to begin by observing that the source density s on the right-hand side of (109) may always be expanded in a Fourier series,

$$s = \sum_i [S_{ci}(r, z) \cos i\phi + S_{si}(r, z) \sin i\phi]. \qquad (110)$$

Here S_{ci} and S_{si} are functions of the r and z coordinates only, the azimuthal variation of $s(r, \phi, z)$ being accounted for in the trigonometric functions. By virtue of equation (105), $u(r, \phi, z)$ may be expressed in a similar fashion, as the Fourier series

$$u = \sum_i [U_{ci}(r, z) \cos i\phi + U_{si}(r, z) \sin i\phi]. \qquad (111)$$

To find values of the unknown coefficients U_{ci} and U_{si}, (110) and (111) may be substituted into (109). Individual Fourier series terms can then be equated individually. A separate problem in the Helmholtz equation is thereby obtained for each term, with the effect that the original three-dimensional problem is replaced by a countably infinite set of two-dimensional problems. For cosine term i, substitution of (110) and (111) into the Helmholtz equation (109) yields

$$\frac{1}{r} \frac{\partial}{\partial r} \left[r \frac{\partial U_{ci}}{\partial r} \right] - i^2 U_{ci} + \frac{\partial^2 U_{ci}}{\partial z^2} = S_{ci}, \qquad (112)$$

and a similar expression applies to the sine terms. These boundary value problems, each comprising one equation like (112) and its associated boundary conditions, can be solved individually and the results combined afterward to produce the complete solution. Of course, the boundary conditions must be similarly separated. Such a separation is always possible if the problem boundary shapes (though not necessarily the boundary condition values) are rotationally uniform, because any function prescribed along an axisymmetric boundary can be given a Fourier series representation analogous to (110) or (111).

An extremely important special case arises in problems with rotationally uniform boundaries when all excitations and boundary conditions possess axial symmetry as well. In such cases, separation of variables becomes attractive and easy, because only the cosine term corresponding to $i = 0$ survives in each series. Thus (112) takes on the particularly simple form

$$\frac{1}{r} \frac{\partial}{\partial r} \frac{\partial U}{\partial r} + \frac{\partial^2 U}{\partial z^2} = S. \tag{113}$$

Subscripts have been omitted, since only one term in the series survives.

Although the axisymmetric and translationally uniform problems both yield to separation of variables, the two-dimensional equations that remain to be solved are not the same. Their computational formulations are correspondingly different, and solution must necessarily be carried out using different numerical techniques. They may be implemented in different solver programs, or may require the existence of multiple options within the same program. These differences in any case explain why the various two-dimensional CAD systems may differ in their capabilities, even though all are two-dimensional.

Axisymmetric Vector Potential Problems

The Laplacian operator in Cartesian coordinates has a similar appearance for both scalar and vector fields, the vector expression being merely the scalar one applied to the x, y, and z components in turn. In other coordinate systems, however, it has quite different forms in the scalar and vector cases. Expressed in cylindrical coordinates, which are the non-Cartesian case of major interest, the Helmholtz operator applied to a vector field $\mathbf{A}$ has the form

$$(\nabla^2 + k^2)\mathbf{A} = \mathbf{1}_r \left[\frac{\partial A_r}{\partial r^2} + \frac{1}{r} \frac{\partial A_r}{\partial r} - \frac{A_r}{r^2} + \frac{1}{r^2} \frac{\partial^2 A}{\partial \phi^2} - \frac{2}{r^2} \frac{\partial A_\phi}{\partial \phi} + \frac{\partial^2 A_r}{\partial z^2} \right]$$

$$+ \mathbf{1}_\phi \left[\frac{1}{r_2} \frac{\partial^2 A_r}{\partial \phi^2} + \frac{2}{r^2} \frac{\partial A_r}{\partial \phi} - \frac{A_\phi}{r^2} + \frac{1}{r} \frac{\partial A_\phi}{\partial r} + \frac{\partial^2 A_\phi}{\partial r^2} + \frac{\partial^2 A_\phi}{\partial z^2} \right]$$

$$+ \mathbf{1}_z \left[\frac{1}{r^2} \frac{\partial^2 A_z}{\partial \phi^2} + \frac{1}{r} \frac{\partial A_z}{\partial r} + \frac{\partial^2 A_z}{\partial r^2} + \frac{\partial^2 A_z}{\partial z^2} \right]$$

$$+ \mathbf{1}_r \, k^2 \, A_r + \mathbf{1}_\phi \, k^2 \, A_\phi \, \mathbf{1}_z \, k^2 \, A_z. \tag{114}$$

Fortunately, it is very rarely necessary to work with the full complexity of this operator. In the vast majority of practical magnetics problems, the vector $\mathbf{A}$ represents the magnetic vector potential, whose divergence is dictated primarily by electromagnetic considerations. Hence the most common situation by far is that of an axisymmetric field, in which **A** has an azimuthal component but no others; and furthermore, that the azimuthal component is invariant with angle. In that case, the general vector Helmholtz operator (114) simplifies to

$$(\nabla^2 + k^2)\mathbf{A} = \mathbf{1}_r\left[-\frac{A_\phi}{r^2} + \frac{1}{r}\frac{\partial^2 A_\phi}{\partial r} + \frac{\partial^2 A_\phi}{\partial r^2} + \frac{\partial^2 A_\phi}{\partial z^2} + k^2 A_\phi \right],(115)$$

which bears some resemblance to (108). Indeed, the resemblance is more than minor; for $m = 1$ the two operators become identical! In other words, the axisymmetric vector potential formulation yields an equation whose form is merely a special case of the axially periodic problem. No separate solution procedures and no separate solving software are therefore required to cope with the different equations.

In the axisymmetric vector problem, as also in the scalar case, numerical instabilities can arise in the process of solution, since the inverse powers of r which appear in (115) create singularities difficult to deal with computationally. They can be avoided by introducing modified potentials V, by setting

$$V = r^{-p} A_\phi .\tag{116}$$

Substitution of the modified potential into (115) yields

$$r^p\frac{\partial^2 V}{\partial r^2} + (2p+1)r^{p-1}\frac{\partial V}{\partial r} + (k^2 r^p - m^2 r^{p-1} + p^2 r^{p-2})V + r^p\frac{\partial^2 V}{\partial z^2} = 0.\tag{117}$$

Finite element analysis of fields requires evaluation of integrals of the form

$$I = \iiint A_\phi \nabla^2 A_\phi \ dr \ rd\phi \ dz.\tag{118}$$

In the axisymmetric case, integration with respect to the azimuthal angle is easily carried out. There results

$$I = 2\pi \iint r A_\phi \nabla^2 A_\phi \ dr \ dz.\tag{119}$$

With substitution of (116) and (117), this integral takes on the detailed appearance

$$I = 2\pi \iint \left[r^{2p+1} V\frac{\partial^2 V}{\partial r^2} + (2p+1) r^{2p} V\frac{\partial V}{\partial r} \right.$$

$$+ (k^2 r^{2p+1} - m^2 r^{2p} + p^2 r^{2p-1}) V^2$$

$$\left. + r^{2p+1} V\frac{\partial^2 V}{\partial z^2} \right] dr \ dz.\tag{120}$$

The lowest exponent of r in this expression is clearly $2p-1$. Hence the integrand in (120) is regular, and no numerical difficulties will arise, provided the index p has *at least* the value *one-half*, i.e., provided the potential in (116) is weighted by at least the square root of the radius. There do exist finite element programs which employ a square-root weighting. Some others use $p = 1$, so as to avoid repeated computation of square roots. From the point of view of numerical stability, there is not much to choose; either will work equally well. However, it is necessary for the user to know which form of weighting was used in solution, so as to be able to recover the true vector or scalar potential for postprocessing operations.

Some finite element programs, particularly early ones, ignore singularities altogether. They avoid singularity problems by using, in each finite element, a single fixed value of r, commonly taken to be the centroidal radius of the element. It goes without saying that this procedure can lead to uncontrolled numerical error, and is best avoided.

In well designed CAD software systems, the actual setting up and solving of equations occupies at most 5–10% of the system software, the vast majority of program code being devoted to the data handling, file management, and human interaction tasks that such systems are required to take care of. Consequently, the need for distinct solvers for the various classes of problem does not imply that all of the CAD system has to be duplicated. With proper design, a major part of the system can remain invariant, only the relevant rather small portion of program code being altered. These remarks must not be construed to imply that equation solving is quickly and easily done; even the most efficient equation solvers take minutes or hours to solve large sets of problems, while the various management tasks typically execute in fractions of a second. In other words, in CAD systems as in many other computing environments, 5% of the program code probably accounts for 95% of the computer resources actually used, and vice versa.

Problem Modelling and Mesh Construction

Setting up magnetics problems for solution by computer involves several tasks. First, the geometric shape of the device to be analyzed must be described to the computer, and a discrete numerical model must be created which satisfactorily approximates the real device. Next, the materials to be used in the analysis must be identified and their properties described. Finally, the boundary conditions and the excitation values must be defined. It is often necessary to compute a solution for each of several closely related cases, for example, to find the magnetic state of a device for various different values of exciting current; therefore, boundary conditions and excitations are usually prescribed several times, one set for each distinct physical case. Once the physical problem has been fully stated in this way, most CAD systems further require the user to stipulate certain purely system-related parameters, such as the type of equation-solving technique to be used or the accuracy levels desired. The entire ensemble of steps needed to set up problems in a form ready for solution is usually termed *preprocessing*. This chapter outlines typical preprocessing steps in CAD systems, omitting only the construction of material property curves which is described elsewhere in this book.

Geometric Modelling and Discretization

A major step in preprocessing consists of describing the geometric shape and size of the object or situation to be analyzed and stating how the object is to be discretized for purposes of analysis. These two activities are sharply separated in some systems, while in others they may merge almost imperceptibly into a single activity. Advocates of separation believe that the discretization ought to be performed automatically or nearly so, with minimal user intervention. Others believe that the user ought to have full control of all approximations involved and ought therefore be allowed as active a role as he wishes in any and every working step.

Finite Element Meshes

In CAD systems based on finite element methods, two distinct approaches are commonly used for generating and storing geometric information about the object to be analyzed. In one approach, a numeric representation of the object itself is produced. In the other, the shape of the object is implied by the finite element mesh which is used in the mathematical analysis to follow.

Finite element methods as mathematical methods are presented in a brief fashion elsewhere, with detailed treatments available in other books specialized to the finite element method. For present purposes, it suffices to say that in the finite element method, every geometric object of interest is viewed as a composite structure made up of more or less standardized parts or *elements*. It is assumed that the field inside any one element is given by a reasonably simple mathematical expression containing a few unknown coefficients. These coefficients are determined by insisting, first, that the fields must satisfy the essential electromagnetic boundary conditions at interelement edges, and secondly, that they must come as close as possible to satisfying Maxwell's field equations over the whole region. Once the constants are determined, the field everywhere in the problem region is known. To find the field at some arbitrary point P, it suffices to evaluate the field expression applicable to the finite element in which P is located, an easy task since the mathematical expressions involved are simple. In practically all currently available electromagnetics CAD software, low-order (linear or quadratic) polynomials are used.

The finite elements most commonly used in current electromagnetics packages are triangular. Quadrilateral elements are also employed, but they are less common. In three dimensions the analogous elements are tetrahedra and regular hexahedra ("squashed bricks"). Both are commonly used. A typical finite element mesh is illustrated in Fig. 1, which shows a loudspeaker magnet assembly. The assembly is axisymmetric, with the axis of symmetry at the bottom edge of the drawing. The loudspeaker voice coil fits into the air gap surrounding the central soft iron cylinder, around the right-hand part of the cylinder as drawn in Fig. 1; the voice coil moves up and down, thereby displacing the loudspeaker cone, which is nonmagnetic and therefore not modelled in the drawing. The permanent magnet material which creates the air-gap flux is cut as a cylindrical ring, and appears in the top half of Fig. 1. Only a minimal amount of air space outside the magnet assembly itself is modelled in this case, because the object of the analysis was to determine the air-gap flux. Were stray flux densities of interest, on the other hand, a much larger air space would need to be modelled.

Finite element methods generally produce "best possible" approximations to the Maxwell field equations, by making the stored energy, total power, or some other globally defined quantity as nearly correct as can be achieved. Designers, however, often need to have not only correct global totals but must also have accurate information about local field values. In

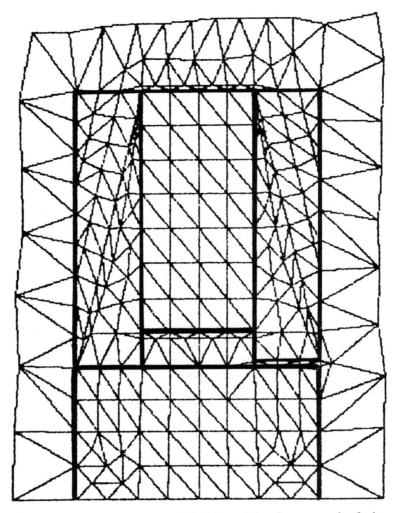

Figure 1. A loudspeaker magnet, modelled by a finite element mesh of triangular elements.

the loudspeaker magnet shown, for example, the total stored energy determines the weight, and therefore the cost, of permanent magnet material. The electrical performance of the loudspeaker, on the other hand, is critically dependent not on weight, but on the level and uniformity of air-gap flux density near the voice coil. High local precision in such sensitive areas is usually achieved by controlling the mesh fineness, using a fairly dense mesh of small elements in such regions, and a coarser mesh in parts where precise field values are not crucial. Such control is necessarily achieved by some form of manual intervention, for there is no automatic way for the CAD system to know what aspects of a problem the designer may think important!

Representation of Finite Elements

All the currently existing magnetic field analysis programs know nothing of geometric modelling and regard a problem region as defined solely in terms of a finite element mesh. The mesh itself is composed of triangular (in some cases, quadrilateral) finite elements. Several existing systems regard meshes as being built up by joining together submeshes, each of which may in turn consist of one or more elements. At the elementary level of small submeshes, elements may be specified explicitly, or implicitly through conventional rules for subdivision of simple primitive shapes such as rectangles.

Most current CAD systems build and maintain data bases in which elements are explicitly specified. To specify a triangular finite element unambiguously, it suffices to give the locations of its vertex points (e.g., by giving the coordinate values of the vertices) and to attach an element label to it. The label serves to identify the element by describing its nature—the type of physical material, source density, etc. It is frequently referred to as an *attribute* label. The vertex points identify the element location and shape. Thus, a triangular element mesh as shown in Fig. 2 is specified by describing each of its elements in turn, by a descriptive statement such as

$$\{(0.0, 0.0), (1.0, 0.0), (1.0, 1.0), A\}.$$

Here the three coordinate pairs specify vertices, while the attribute label A describes the nature of the element. Attribute labels are used to identify materials or current densities, or simply to label particular regions of a problem so they can be picked out for attention afterward. Labels as used in various CAD systems may be integer numbers, single printable characters, or names composed of alphanumeric characters. In MagNet, for example, any one upper-case letter, numeral, or special character may be used, with the exception of the underscore character and the blank space. Labels used by the MAGGY system, on the other hand, are multicharacter words, such as *IRON, AIR,* or *Z217.* A few systems employ multiple labels, allowing materials and excitations to be handled separately.

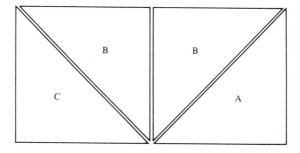

Figure 2. A simple triangular finite element mesh.

While constructing a finite element mesh, it is not important to state explicitly what material properties or sources are associated with each element. It is only necessary to tag each element with its attribute label so that all elements belonging to a common grouping—all elements made of the same magnetic material, for example—carry similar labels. The labels themselves are abstract, in the sense that their meanings are divorced from their form, and will be stated later. In fact, abstract labelling of this kind can make it very easy to define sequences of closely related problems, since it is only necessary to redefine the meaning of a label in order to alter current densities or material characteristics.

In most finite element models every vertex point has several elements touching it. In fact, the total number of elements is often larger than the number of points. A considerable reduction in data volume is achievable if the point coordinates are stated only once, in a separate list, and elements are defined by giving point numbers. The above element might then be represented by

$$\{1, 3, 2, A\},$$

where the points are specified in an entirely separate list as

1:    *(0.0, 0.0)*
2:    *(1.0, 1.0)*
3:    *(1.0, 0.0)*
4:    *(. . . . .*

This distinction in internal representation may not seem very important to the user, but it is significant in one respect. In the first representation given above, points only exist in association with finite elements; there is no such separate entity as a "point", and none can be made to exist except by describing elements. In the second representation, a point can be permitted to exist even if there are no finite elements attached to it. This difference normally has a strong influence on the system command language and its rules: if points exist as separate entities, there must exist commands for defining them, but if points exist only in association with elements, then the command language cannot contain any instructions that deal with points. Similar considerations naturally apply just as well to other data entities in CAD: what isn't there cannot be controlled.

Representation of Boundary Conditions

Magnetic fields in principle occupy all of space, while computational analysis normally deals only with finite regions. These regions are ordinarily defined by boundaries, which may be part of the problem in a natural way. Such boundaries are formed, for example, by planes of symmetry. In other cases, boundaries may be purely artificial, having been introduced as an analytic convenience.

Finite element solution techniques generally produce solutions which are expressed in terms of potential values at the element nodal points. Boundary conditions are therefore very conveniently described in terms of nodal potentials, rather than as conditions along element edges. Two forms of condition are commonly encountered in practice: fixed potential values and relationships between potential values at distinct points. The former type most often takes the simple form

$$A_K = c, \tag{1}$$

meaning that the potential A_K at nodal point K is required to have the given constant value c. Because such a condition involves only the potential at a single node, itis frequently referred to as a *unary* constraint condition.

Relationships between potentials at two distinct points, say J and K, are referred to as *binary* constraints. They typically require equality of potentials at two points,

$$A_K = A_J. \tag{2}$$

Equality requirements arise in periodic or repetitive structures, where potential values are known to repeat at prescribed locations. For example, the fields in a normal electric machine are known to repeat every 360 electrical degrees (i.e., every pole pitch). In less simple cases, binary constraints may take a more general form, such as

$$f(A_K, A_J, W) = c_W, \tag{3}$$

where $f(a, b, W)$ is some function of the variables a, b. The symbol W here is an abstract label, which serves to identify a group of boundary constraints of like character. It is abstract in the same sense as an element label: the character W itself has no predefined significance, so that it will be necessary to define its meaning elsewhere. In most practical design problems, only a few distinct types of relationship occur although many points may be constrained. It is therefore convenient to use abstract labels to classify relationships, and to assign numeric values by defining and redefining the label meanings rather than by attaching actual numerical values to the element nodes. Curiously, the latter approach is often used in stress analysis packages.

In the MagNet system, the two common types of constraint condition are given a single formal representation. A binary (two-potential) constraint, e.g., one that relates the potential at node *137* to that at node *831*, is represented internally within the system in a manner reminiscent of equation (3), as an ordered pair *137 831* of node numbers, with an associated labelling character:

$$[137\ 831\ W].$$

The simpler unary (single-node) constraint, which only deals with the potential at node *516* (say), is represented as a binary relationship to node zero, a node number which cannot exist if nodes are numbered beginning with *1*:

$$[516 \; 0 \; S].$$

Here W and S are arbitrary labels, which identify both the functional relationship $f(a, b, W)$ of equation (3) and the right-hand side c_W of that equation. These internal representations are of course invisible to the system user, who neither knows the numbers of individual points nor cares to know them. But just as the way elements are stored internally affects the language with which the user communicates with the system, so does the internal storage form of constraints. To impose a condition on a pair of nodes, the user is thus obliged to (1) declare whether the condition is unary or binary, (2) identify one or two points, and (3) specify a label. As indicated in more detail below, identification is usually achieved by pointing at nodes. Definition of the label meanings is deferred till later, in much the same way as the meanings of attribute labels associated with elements. This approach permits sequences of problems, differing only in their boundary conditions, to be defined by giving a variety of label definitions.

Several of the currently available CAD systems for magnetics include lengthy sets of symmetry conditions, permitting odd and even symmetries in various combinations of symmetry planes. Symmetry conditions amount to particular forms of boundary conditions, and are therefore special cases of the unary or binary conditions given above. Whether it is better to design systems which permit the user to specify such conditions in a direct but somewhat abstract form (as above), or to allow them to be specified through special descriptive statements such as symmetry commands, is a matter of taste. As a general rule, the more abstract forms are few and brief, while the physically oriented descriptions are more obviously directed at the problem. However, since they must needs be directed at special cases there are generally large numbers of them.

Mesh Models for Geometric Objects

A complete model of a problem to be analyzed must contain full information about the geometric shapes and materials involved, as well as all the boundary constraints to be satisfied and the sources to be included. It must therefore include at least a set of elements and a set of constraints. Since both need to be attached to points, a set of point coordinates is also required. It is useful to append to these minimal data a string of alphabetic text specified by the user. This text might include, for example, a few descriptive words about the model, its date of creation, name of the author, and perhaps other identifying details. A complete finite element model of some geometric structure thus comprises

a list of elements, e.g.,	[(2, 7, 12, L), ...]
a list of constraints, e.g.,	[(12, 18, &), ...]
a list of points, e.g.,	[(0.5, 0.72), ...]
a list of text characters, e.g.,	[THESE ARE WORDS]

When needed, the lists that make up a model are found through a directory of models. Each directory listing consists of a model name and four address pointers which show where the four lists of that model are filed.

The four lists which describe a model are interdependent and form the pointer-linked data structure shown in Fig. 3. The element list and the constraint list contain pointers to the coordinate list. Hence points may exist independently of elements in this data structure, but elements can only be made to exist if their vertex points exist. The text is of course entirely separate from the lists that give geometric and topological information.

Label definitions do not form part of the geometric description, but are introduced separately, as discussed further below. Their significance is electromagnetic, not geometric or topological. This separation of geometric and electromagnetic data, achieved through the use of abstract labels, enhances working convenience substantially. For example, extensive alterations can be made in materials, sources, or boundary conditions, without in any way altering the (often large) geometric data set.

Every CAD system must allow the user facilities for creating, inspecting, and possibly modifying a describing data structure like that in Fig. 3, or one equivalent to it. Fully interactive systems permit direct graphical display manipulation, so that the creation, inspection, and alteration merge into a single *editing* process. Systems oriented to batch processing use formal command languages for specifying the data to be included in

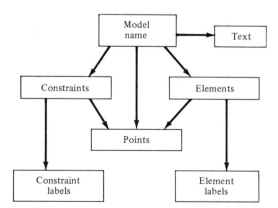

Figure 3. Relationship of data lists in a model. The element and constraint lists only contain topological, not geometric, information

the structure, and they usually incorporate some internal checking to spot at least the most common data blunders. Data specification invariably involves a lot of manual work and is therefore prone to error. To simplify error detection, practically all systems provide some form of graphic echo prior to problem solution, in accordance with the ancient adage of all computing: *garbage in produces garbage out*.

Mesh Generation and Editing

In interactive CAD systems, the creation or modification of geometric models and meshes becomes a matter of *editing* the underlying data structure, i.e., of adding, removing, or modifying its parts until the data structure adequately describes the object to be analyzed. The editing programs now in use follow the general principles common to text editors, which are familiar to most computer users. Interactive editors were not common in CAD until about 1978 or 1980, because the necessary computer graphics support was expensive and difficult to provide. Since that time, however, interactive computing has become the norm, and few if any new batch-oriented systems have been created.

In this section, the major operations required to edit a problem data base are described briefly, using the MagNet system as a specific example. MagNet is the most strongly interactive system currently available and is therefore an appropriate choice to illustrate interactive operations.

Mesh Editing

Mesh editing programs are programs able to modify mesh models, that is, to add points, elements, constraints or text characters, or to remove them. While details differ, editing programs must generally contain modification operations which address the four types of entity individually, through four or more basic editing commands. For a program to qualify as a true editor, these operations must always produce valid models from valid models, for only in this way will the system structure impose no particular sequence on the editing operations. This point is critically important: the very essence of interactive computing is the freedom to vary the sequence of operations, to have second thoughts about actions already performed, and to go back to alter what is already there. In batch operations, on the other hand, there is no particular objection to being constrained to a specified sequence of data and operations.

The MagMesh subsystem of MagNet is a mesh editor strongly oriented to interactive graphics, using a raster-refreshed (television-type) display screen and a graphics tablet, as in Fig. 4, for graphical communication. In addition, a standard alphanumeric terminal keyboard is used for alphabetic input. Whenever the system is ready to receive user commands from the keyboard, it so signals by sending the prompting message

```
>> NAME ready.    Command:    _
```

Here NAME is always the name of the model currently being edited.

Most command sequences used in MagMesh are initiated through the keyboard. But wherever communication between user and system refers to pictorial or graphical rather than verbal entities, graphic communication, through the graphics screen and graphics tablet, is used because it is more natural by far. When the system is ready to receive graphic input, a crosshair cursor is shown on the screen, and the user may "point" to parts of the screen by moving the cursor. The cursor shown on the screen instantaneously tracks the location of the graphic input stylus (or puck) on the tablet. Especially when a pen-like stylus is used as the cursor control device, as in Fig. 4, the user rapidly gains the impression that he is actually drawing on the screen, rather than on the tablet!

There are four basic editing commands in MagMesh, corresponding to the four basic data entities in the mesh. An existing model may have *points, elements, constraints,* or *text* added, altered, or removed through operations initiated by the keyboard commands CONStraint, ELEMent, POINt, and TEXT. These are followed by information on whether replacement, addition, or deletion of list items is required, and what the new items (if any) are.

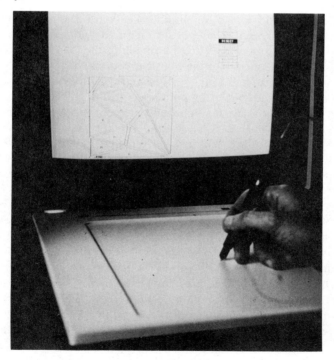

Figure 4. Working with the graphics tablet and screen, the user has the impression of actually writing on the screen itself.

Element editing operations comprise element *definition* (i.e., adding elements to the model), element *deletion*, and element *label alteration*. Issuing the ELEM*ent* command at the keyboard produces a menu on the graphic display screen, the possible menu selections corresponding to these activities. (Such menus appear in various illustrations throughout this book, for example in Fig. 7, further below.) To define new elements, DEFINE is chosen from the menu by a cursor hit on that word, i.e., by moving the crosshair cursor to that word and pressing down the stylus on the tablet. The new element to be added is described by an ordered sequence of three cursor hits on the three points which define the triangle. The points are highlighted in the graphic display as they are identified, and the element is drawn on the screen as soon as its definition is complete. Errors can be corrected immediately, by repeating the last cursor hit (or hits). To delete an element, it suffices to select DELETE on the menu, then to point at the element in question. A cursor hit anywhere within the element suffices for this purpose. In a similar way, the LABEL selection allows alteration of element labels.

Throughout all activity under the ELEM*ent* command, the graphic menu remains on the screen; to switch from one action to another, it suffices to point at the appropriate menu item. Any sequence of definitions, label alterations, and deletions is valid, and continues until QUIT is selected on the menu. The QUIT selection terminates activity under ELEM*ent* and causes MagMesh to prompt for another keyboard command. The MOVE selection allows the user to move the menu to another screen location. Constraint editing is accomplished by invoking the CONS*traint* command, which is similar in principle to the ELEM*ent* command.

In all operations under MagNet, geometric entities such as points or elements are always identified by pointing with the graphic stylus and cursor, never by keyboard input. Point and element numbers are not shown on the screen, nor indeed is there any way for the user to have them displayed. Operations are thus always directed to "this point" or "this element", but never to "element number 559". In other words, the pointing action becomes part of the communication vocabulary, just as a pointing finger accompanies the words "I will take this fish, please" in everyday speech. Pointing is used as a *selection* tool, e.g., to identify three points by selecting them from a much larger set displayed on the screen, or by selecting items from a set of menu choices shown on the screen. Pointing is occasionally also used as a *specification* tool to communicate position or length; but such purposes are usually better served by a numeric keyboard. This arrangement requires both reasonably fast communication speeds and the ability to erase parts of the screen display selectively (e.g., to put up and then remove menus). It contrasts sharply with semi-interactive systems designed to be operated with storage-tube terminals or paper plotters. Such systems generally display node and element numbers on the screen, and the user identifies individual parts of the model by copying their numbers at the keyboard.

The POIN*t* command allows points to be inserted in a model or to be deleted from it. Points are defined by entering their coordinates through the keyboard, or by means of cursor hits. Provision of keyboard input is essential. Point locations must at times be specified to as many as five or six significant figures, a precision feasible only through keyboard entry of numbers. Furthermore, at least the first two points defined in any model must be entered through the keyboard in order to define screen scale. Graphic displays in MagMesh are automatically scaled so as to fill the screen, an impossible task until at least two distinct points exist in a model. On the other hand, DELE*tion* of points is best done through graphic selection by pointing. When points are deleted, any elements or constraints associated with the point to be deleted are deleted also.

The model descriptive text may be displayed, and may be replaced by new text, through the TEXT command. This simple operation is of great practical value. User files can and do grow quickly as variations on designs are produced, with dozens of models the result. Tagging each with a descriptive phrase is by far the best way of keeping track of them all.

Mesh Construction: a Simple Example

To illustrate the technique of mesh construction with an interactive editing system, consider the problem of finding the leakage inductance of a small transformer, as shown in Fig. 5. For simplicity, the transformer is assumed to extend infinitely into the paper and to have a core of infinitely permeable material. To analyze the leakage field, it suffices to model one-quarter of the window space of the transformer. Its corners are at *(0.,0.)*, *(0.,1.5)*, *(1.,0.)*, *(1.,1.5)*, while the winding to be modelled occupies the rectangular space bounded by *(0.0,0.0)*, *(0.0,1.3)*, *(0.7,0.0)*, *(0.7,1.3)*. These seven points define implicitly the minimal possible triangulation of the region to be analyzed.

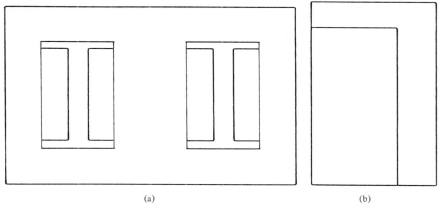

(a)   (b)

Figure 5. (a) A simple transformer, with core assumed infinitely permeable for calculation of leakage inductance. (b) One-quarter of the transformer winding window, the minimal region needed for analysis.

The finite element model of the quarter window is created by initially creating a new model called (say) LEAK, by asking LEAK to be made active for editing (instead of whatever model was active last). This request is communicated by the MODE*l* command. For example,

>> YECH ready. Command: MODEL LEAK

The system replies by requesting the user to enter the model describing text, say MINIMAL MODEL OF QUARTER WINDOW. A model called LEAK, composed of the text segment only, is thereby created. It contains no points, no elements, and no constraints. Although it is a formally valid model (it comprises the necessary lists, albeit of zero length), it is not the desired model, and must be made into the correct one by editing. This situation is typical of editing-based systems: new models are always created as empty models, then edited (by adding points, elements, and constraints) to give them the desired geometric content. An empty model, be it noted, is a formally proper model, not at all the same thing as the absence of a model. To turn the empty model LEAK into one useful for the transformer problem, the minimally necessary fiducial points are entered:

>> LEAK ready.   Command: POINT 0   0
                              Point = 1.   1.5
                              Point = 1   0
                                . . . . .
                              Point = 0.7   1.3
                              Point = QUIT

In the above, and the following, all words entered by the user appear in capital letters only, while the system responses and prompts are shown in upper and lower case as appropriate. Thus *POINT* in the above is a command entered by the user, while *Point* is a system prompt. Because points are most often entered in sets, not one at a time, MagMesh keeps prompting the user for additional points until he signals (by entering QUIT) that he wishes to cease.

Graphic display of the points becomes possible as soon as the model contains two points, enough to establish screen scale. At that time, any previous display on the screen is cleared and the first two points are plotted. Any subsequent points are added to the display as they are defined.

Having defined the minimally necessary points, the user issues the ELEM*ent* editing command

>> LEAK ready.   Command: ELEMENT

The graphic cursor appears on the screen, along with a screen menu. Selecting DEFINE on the menu, the user proceeds to indicate triples of points, each of which defines a triangular element. The points are selected by means of cursor hits, that is, by moving the graphic stylus and crosshair cursor to the desired point, then depressing the stylus. Point numbers are neither needed nor even known to the user. Initially, it suffices to produce a minimal triangulation, remembering to label the air space and the winding with distinctive labels, say A and W respectively. The result is then as shown in Fig. 6.

Mesh refinement may be introduced in several ways. The most direct method is provided by the ELEM*ent* command itself, through the BISECT and TRISECT menu selections. BISECT accepts a cursor hit near an element edge, then finds the nearest point on that edge. It creates a new node at that point, and replaces the two triangles which originally abutted on the edge by four new ones. In a similar fashion, TRISECT accepts a cursor hit inside a triangle, creates a new node at that point, and replaces the original triangle by three new ones. The new triangles carry the same identifying labels as the elements they replace. The actions of BISECT and TRISECT are illustrated in Fig. 7.

Operations such as bisection and trisection can be made somewhat more sophisticated. For example, some editing systems examine all the elements in the immediate neighborhood of any bisected elements and rearrange other adjoining elements so as to make individual triangles as nearly equilateral as possible. In general, such operations are possible only with triangular elements (or tetrahedral elements in the three-dimensional case), because the mathematical theory underlying the necessary computation techniques has not been developed for quadrilaterals and hexahedra.

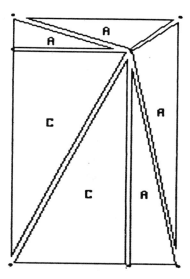

Figure 6. Minimal triangulation of the quarter window region of a transformer, for leakage computations.

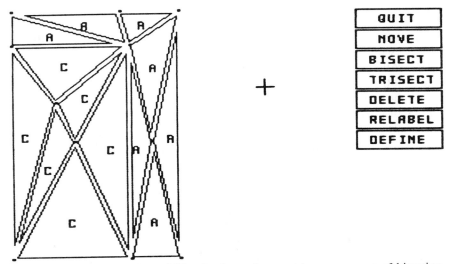

Figure 7. The triangulation of Fig. 6, after refinement by a sequence of bisection and trisection operations.

By successive refinement of coarse initial models, quite complicated finite element meshes can be produced in a short time. It should be noted that numeric coordinate values need to be entered through the keyboard only for leading dimensions, i.e., for points which must be accurately placed in order to define the geometric shape and size of the region. Any points newly generated in the process of mesh refinement are assigned locations automatically.

An alternative approach for getting a new model started under Mag-Mesh is to allow the system to do most of the work, by using the MAKE command. This command combines the MODE*l* and POIN*t* commands with a measure of automatic mesh generation. Having invoked MAKE, the user identifies the set of points that defines the perimeter of an area, in a fashion analogous to the POIN*t* command. The system then produces a subdivision into elements, using the smallest number of elements possible, and (so far as there is any choice) using as nearly equilateral triangles as possible. It is then up to the user to decide how the mesh should be further refined. MAKE is the most common command used for initiating new models within the MagMesh subsystem. In addition to accepting newly defined points, it also allows transfer of a point set from one model to another, so that boundary shapes can be matched exactly in different models. Of course, MAKE cannot do anything which the four basic editing commands cannot do; but it is often valuable as a time-saver.

Model Transformation

The editing operations above change a model by altering it topologically, through removal or addition of parts. A model may also be transformed through operations which change it geometrically by modifying the locations of its nodal points but leave it topologically invariant. There are four basic model transformations of this type, and the MagMesh system provides one command for each. These are MIRRor, ROTAte, SCALe, and SHIFt. All four perform coordinate transformations which map a model into another of similar topology but different dimensions, shape, or position. The element and constraint lists of the model, however, are left unaltered. In other words, each of these operations transforms all the coordinate values stored in the point list. The four operations are fundamental, in the sense that each produces an effect which cannot be obtained by any combination of the other three. Some editing systems may also provide additional commands, which are combinations of the above. In MagMesh, such combinations can be defined by the user, and added to the system, as indicated further below.

MIRRor is a reflection transformation, which alters coordinates so as to produce an image of the model reflected about a certain straight line. The line may be specified numerically through the keyboard, or graphically by means of two cursor hits.

The ROTAte command rotates the model by a specified angle in the range 0–360 degrees, about a specified point. Again, the arguments may be communicated numerically through the keyboard or by pointing at the graphic display. For example,

ROTA 45 13.5 0

means: *rotate the entire model 45 degrees counterclockwise about the point (13.5, 0.0).*

SCALe does just what one might expect: it multiplies the x and y coordinates of all model points by two constant factors, which may (but need not) be equal. The factors may be given numerically through the keyboard, or they may be implicitly specified by giving two vectors. In the latter case, the scaling factors are computed so as to map the first vector into the second.

The SHIFt command translates all model points by a translation vector which may again be specified either by the keyboard command or by graphic means. For example,

SHIF 2.5 −2.5

translates the entire model in a southeasterly direction: 2.5 units to the right and 2.5 units downward. The alternative graphic specification would consist of two cursor hits, to specify the beginning and ending points of the translation vector.

After every transformation command, the model as transformed is presented on the display screen, unless displaying is deliberately suppressed by the user. All display operations translate and scale screen coordinates so as to have the plot fill the screen.

Model transformations are useful for effecting geometric alterations on meshes. However, their real importance only comes to the fore when coupled with set operations, as detailed below.

Multi-Model Operations and Model Libraries

Models of complex devices are often best created by joining models of their individual parts, just as the physical devices themselves are fabricated by making parts and joining them. Each partial model may correspond to an actual manufacturable part, such as a shaft, or to an identifiable segment of a regular structure, such as a tooth pitch in an electric machine rotor. Joining parts together presupposes that the parts are separately created and stored in a model library, a bin of parts, as it were. Effective geometric model building thus requires two ingredients: the ability to join two geometric models and the ability to manage a library of models.

Set Operations

In geometric modellers used for automated drafting and mechanical design, set operations are fundamental to system operation. Intersections, unions, and exclusions of valid geometric objects are variously provided. In mesh editing systems such as MagMesh, only one true set operation, the union, is normally essential, for any other requirements can be handled through the elementary editing operations in the rare cases when a need for them does arise. The union operation is invoked in MagMesh by the JOIN command, which replaces the model currently active with the union of the active model and another model named in the command itself. For example,

$$\text{>> FRED ready. Command: JOIN NAME}$$

creates a new active model FRED, whose point set is the union of the point sets of NAME and the old model FRED; its set of elements is the union of the element sets of NAME and the erstwhile FRED; and its set of constraints is similarly the union of the two earlier sets.

Joining two models is a fairly complicated operation, for the models to be joined must satisfy a number of topological compatibility checks. Constraints must not conflict, that is, a point constrained in one model must not be differently constrained by the other. Similarly, elements which occur in both models must have matching labels. Of course, no overlapping or intersecting elements should be created by the union. Complete checking of this sort is more or less feasible, but it can become very time-consuming. Extensive and detailed validation is generally considered essential in batch-run programs, but in an interactive system such as MagMesh some checking is usually sacrificed in favor of shortened response time. In most cases, a mere glance at the display screen will do as much consistency checking as many millions of machine operations; the human eye is surprisingly good at spotting such anomalies as overlapping elements.

Construction of Complex Models

The ability to transform models and to form unions lends the user great power when dealing with repetitive or quasi-repetitive structures. Typically, a structurally regular model will be built up by first constructing a small portion or elementary cell of the overall structure. This portion is replicated the requisite number of times. The various copies are transformed and edited as required, then joined to form the larger structure. In many applications, the larger structure will in its turn form an elementary cell of the model to be ultimately analyzed and will thus undergo further replication and modification.

As an example of such model construction, Fig. 8(a) shows a finite element model of half a slot pitch of an electric machine stator. This model was constructed using the methods outlined above: a few key points of the shape were entered through the keyboard, a mesh of triangles was constructed, and the final model was arrived at through a sequence of mesh refinement operations. To make up a full slot pitch, as in Fig. 8(b), a temporary copy is first made of the half slot pitch, using the COPY command which replicates the currently active model:

>> SLOT ready.   Command: COPY TEMP

The model TEMP now contains a copy of the original SLOT; TEMP becomes the active model. A mirror image of TEMP is made by issuing the MIRRor command,

>> TEMP ready.   Command: MIRROR

When the graphic cursor appears, two points on the mirroring line (the tooth centerline) are identified so as to define the mirroring line. Joining, and subsequent deletion of the unneeded temporary copy, are effected by

>> TEMP ready.   Command: MODEL SLOT
>> SLOT ready.   Command: JOIN TEMP
>> SLOT ready.   Command: DELETE TEMP

The result is the model of a full slot pitch, Fig. 8(b). To model two slot pitches as shown in Fig. 8(c), a similar procedure is followed. A copy of the single slot-pitch model is made, rotated one slot pitch, and joined to the previous one. One possible command conversation which will accomplish this objective is

```
>> SLOT ready.    Command:    COPY TEMP
>> TEMP ready.    Command:    ROTATE
>> TEMP ready.    Command:    MODEL SLOT
>> SLOT ready.    Command:    JOIN TEMP
>> SLOT ready.    Command:    DELETE TEMP
```

When the ROTA*te* command is accepted by the system, the crosshair cursor
is presented on the screen, and the user is expected to define the angle of
rotation by graphic input. He does so by means of three cursor hits: two
to identify the center of rotation and a ray pointing outward from it, and
a third to specify the direction the ray is to point after the transformation
has been completed. In the present case, the center of rotation is the
center of the stator itself, while the two ray directions are the directions of
the two edges of the full slot-pitch model.

   A full model of the machine stator may next be obtained by continuing
the sequence of rotations and joinings. Doing so obviously involves a
good deal of repetitive work, so it is quite often a good idea to program
the operations for automatic repetition, as discussed in a separate section
of this chapter.

   An important point to observe is that replication of the geometric
shapes brings with it replication of element labels and constraint labels (if
there are any). It will usually be necessary to perform some amount of
label editing on the finished stator model, for example, to allow windings
in the various slots or slot groups to carry different currents.

   To analyze the performance of an electric machine, a model spanning at
least half a pole pitch is generally required. Using the techniques
described above, separate models can be built of the machine rotor and
stator. These are often quickly and easily made, since much repetitive
structure is likely to be involved. The two are then joined, as shown in

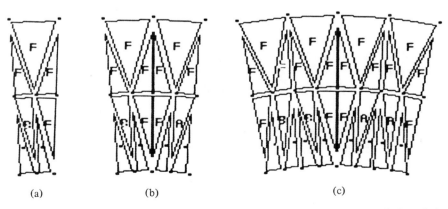

(a)   (b)   (c)

Figure 8. (a) Half a slot pitch of an electric machine stator. (b) The full slot pitch
model, arrived at by mirroring and joining two halves. (c) Model of two slot
pitches, obtained from Fig. 8(b) by replication and union.

Fig. 9(a). However, the resulting model consists of two disjoint pieces, since neither the rotor nor the stator model contains any details of the air gap. The missing portion can be created as a separate model, which possesses the same boundary points as the rotor and stator together but has all its elements within the intervening air space. In MagMesh, such tasks are effectively handled by the MAKE command, which permits the user to make a new mesh, borrowing a set of outline points from an existing one. If the command

>> MACH ready.   Command: MAKE GAP

is issued, the crosshair cursor is presented on the graphic display screen, and the user is expected to identify a sequence of points which form the outline of a new problem region. A model of the air gap is made by identifying the air gap outline by means of cursor hits; as the work proceeds, a "rubber band" polygon is drawn on the screen showing the outline as given by the points specified so far. When the outline is fully defined, the enclosed region is automatically triangulated and left active for further editing if desired. The outline as constructed on the screen appears in Fig. 9(a), and the result of subsequent automatic triangulation is shown in Fig. 9(b); it should be noted that the display is scaled so as to have the gap model fill the available screen space. To complete the full electric machine model, it is only necessary to form the union of rotor, stator, and air gap by means of the JOIN command.

Model Library Management

Duplication, transformation, and joining of models only makes good sense if several models can be retained in the CAD system concurrently. In other words, any good mesh editor must be backed up by library facilities which permit storage and subsequent access.

From the magnetic device designer's point of view, a library of geometric models is a collection so organized as to permit retrieval of individual models at will. (From a computer scientist's point of view, it is a random-access file with an internal directory structure). In principle, such a library is similar to a library of magnetic material characteristics. That is to say, the various models in the library are filed in accordance with some predetermined scheme, and a directory is maintained to show what models the file currently contains. Every model is assigned a name, and the directory correlates names with locations of the corresponding models on the physical storage medium, which is usually a magnetic disk.

Management of a model library requires at least the ability to create, to destroy, and to copy models. This need is satisfied by the commands DELEte, DIREctory, MODEl, and COPY. Two of these, MODEl and COPY, have already been mentioned briefly.

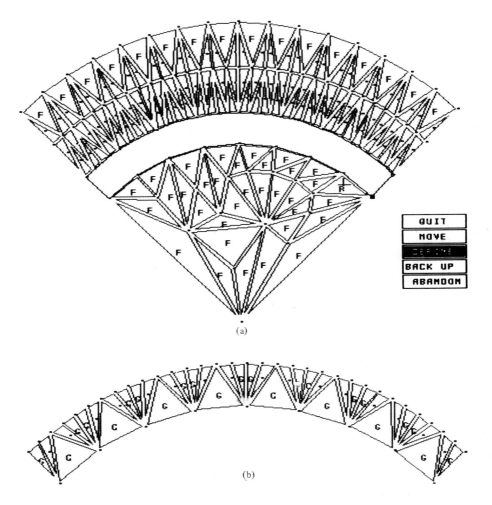

(a)

(b)

Figure 9. (a) Model of electric machine rotor and stator; the menu visible on the screen is part of the MAKE command. (b) Minimal triangulation of the air gap, obtained from the model of Fig. 9(a). The scale is altered in the display so as to fill the screen.

In MagMesh, one model always occupies an *active* role, the others remain passively stored in the library. A model stored in the library may be made active by means of the MODE*l* command. The same command serves to create a new model; if its name does not yet exist in the library, an empty model is created and made active. The empty model contains no points, elements, or constraints, but it does have a directory entry and it is allocated storage space; it may be subsequently edited. The DELE*te* command has the obvious and opposite effect: a model is deleted from the library and ceases to exist thereafter.

The COPY command differs a little from MODE*l* and DELE*te*, in that both the model to be copied and the name to be assigned to the copy must be identified. In some systems, commands equivalent to COPY therefore involve two names, e.g., COPY OLD NEW. In MagMesh, only the new name is shown in the command, the model to be copied is always understood to be the currently active model.

The DIRE*ctory* command produces a listing of the names of all currently filed models and indicates their sizes, in terms of the numbers of nodal points and elements. The latter form of presentation is preferable to giving computer-related information such as file block numbers, for it relates to the user's experience with models rather than to the characteristics of any particular computing system.

The ideal size of a model library should be small enough to permit its contents to be examined conveniently and easily and to allow individual items to be retrieved quickly. On the other hand, it should be large enough to allow all models related to a single project to be housed in a single library. A library capable of housing several maximal-sized models, as well as a few dozen smaller models of parts, is generally adequate.

Batch and Interactive CAD Systems

While interactive computing becomes more popular every day, batch-oriented analysis programs remain in wide use as well. The use of interactive software systems is still impeded by imperfect standardization of computing hardware; but batch-run programs typically assume only the minimal input–output facilities possessed by virtually any computer.

In batch processing, *all* the programs, processing instructions, and data that make up the problem run must be made available to the computer system as a single package at the beginning of the run. This single fact shapes the entire working pattern and the CAD system structure into a mold totally different from interactive computing. Processing instructions—details of meshes, magnetic material descriptions, indeed everything else—are assembled initially into a file, which is read by the CAD system without user intervention. There is no room for second thoughts, no chance of altering the sequence of processing once the job is launched. Small errors can lead to bitter disappointment, but there is a worthwhile *quid pro quo*: the user can assemble the input file at a time and place away from the computer. Only brief, limited access to the computer is actually required by the user, so that batch-run programs are ideal in situations where CAD work must perforce be done on a large multi-user machine.

Problem Setup in MAGGY

Of the CAD systems now generally available to the electromagnetics community, MAGGY is by far the most strongly oriented to batch work, just as

MagNet is the most interactive. It is therefore used here to illustrate how problems are set up in a batch-processing environment.

Input to MAGGY is taken from any device functionally equivalent to a punched-card reader, i.e., any device capable of sequentially reading an input stream of text characters but unable to back up. Output is directed to a printer-like device, one capable of absorbing a sequential stream of text characters. In practice, disk files are normally used as input and output devices; in principle, magnetic tape or punched cards could be employed instead. There is also provision for generating an output tape for off-line graphic plotting, using a pen and ink plotter. These hardware requirements are easily met by the great majority of computing installations.

MAGGY differs from MagNet in several other respects unrelated to batch processing and furnishes an interesting contrast for that reason. Where MagNet concentrates on triangular finite elements with their great geometric flexibility, MAGGY employs quadrilateral elements. The quadrilaterals need not be rectangular, so that the variety of geometric shapes which can be built is still quite broad.

To solve a problem, the MAGGY user creates an input file which contains the geometric details, material properties, current densities, symmetry or boundary conditions, mathematical methods to be used, in fact *everything* the system is to know or do, including output requirements such as what quantities to plot and how. This input is placed in a text file as a set of *blocks*, groupings of related command statements. Each block deals with a different aspect of the task and therefore contains statements different in type and kind. For example, there is a *curve block* in which material property curves are defined, an *algorithm block* which defines the finite element mesh as well as details of the mathematical methods to be used, a *print block* which contains instructions about the results to be printed out, and so on.

Mesh Specification

Meshes used by the MAGGY system are always topologically rectangular, although their geometric form is flexible. They are "rubber meshes"; that is, every mesh used by MAGGY may be regarded as a mapping into the x-y or r-z plane of a rectangle in the u-v plane, in which the mesh lines are straight and parallel. This rectangle is taken to be n_u by n_v mesh lines in extent, as illustrated in Fig. 10. Since every mesh used by MAGGY is a mapping of such a rectangular mesh, description of any geometric entity in terms of its mesh representation must consist of two parts: (1) a statement of how many mesh lines are required, i.e., the values of n_u and n_v, and (2) a statement of the coordinate transformation rules by which the rectangle in the topological reference plane (the u-v plane) is mapped into the problem plane (x-y or r-z). The transformation specification is obviously the more complicated of the two.

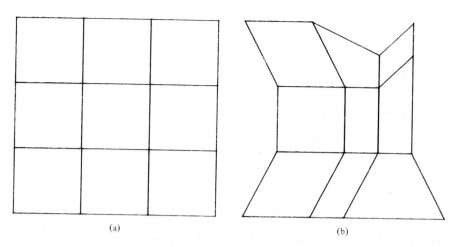

(a)   (b)

Figure 10. (a) Rectangular region of the u-v plane, with n_u = 4 and n_v = 4. (b) A mapping of the same region into the x-y plane.

Mesh transformation rules in MAGGY are stated in a descriptive fashion, by means of mesh statements. Each such statement specifies the coordinate values in the problem (x-y or r-z) plane which are traversed when following the corresponding mesh line in the topological reference (u-v) plane. For example,

$$X(1) = 0.0, \ 0.5, \ 0.5, \ 0.0$$

is read: "The x-coordinates at mesh points along the first mesh line, $u = 1$, are 0.0, 0.5, 0.5, 0.0". This statement in fact describes the x-coordinate placement of the leftmost line in Fig. 10(b). Similarly,

$$Y(1) = 0.0, \ 0.0, \ 0.0, \ 0.0$$

specifies that the first mesh line in the horizontal direction is straight, a segment of the x axis. Continuing, a complete description of the mesh in Fig. 10(b) is given by

$$
\begin{aligned}
X(1) &= 0.0, \ 0.5, \ 0.5, \ 0.0 \\
X(2) &= 1.0, \ 1.5, \ 1.5, \ 1.0 \\
X(3) &= 1.5, \ 2.0, \ 2.0, \ 2.0 \\
X(4) &= 3.0, \ 2.5, \ 2.5, \ 2.5 \\
Y(1) &= 0.0, \ 0.0, \ 0.0, \ 0.0 \\
Y(2) &= 1.0, \ 1.0, \ 1.0, \ 1.0 \\
Y(3) &= 2.0, \ 2.0, \ 2.0, \ 2.5 \\
Y(4) &= 3.0, \ 3.0, \ 2.5, \ 3.0
\end{aligned}
$$

In complicated meshes, such descriptions can become quite lengthy. The MAGGY input language MAGLAN therefore permits a large variety of

abbreviated forms. For example, it is permissible to write *2*0.* instead of
0.0, 0.0, so as to save effort.

Automatic mesh refinement is provided for in MAGGY in an implicit
way. Any mesh lines of the *u-v* plane for which no transformation rule is
prescribed are transformed by linear interpolation between the nearest
specified lines. For example, in the statement group

$$X(1) = 5*0.0$$
$$X(2) = 5*1.0$$
$$X(3) = 5*2.0$$

the second line is redundant (though it does no harm). Leaving *X(2)*
undefined, and specifying merely

$$X(1) = 5*0.0$$
$$X(3) = 5*2.0$$

causes the undefined mesh line to be assigned coordinate values by interpolation between the corresponding values in *X(1)* and *X(3)*, yielding
exactly the same values as entered explicitly in the fuller description
above.

Topologically regular meshes are used by quite a few CAD programs,
especially early ones, either to simplify mathematical computations or to
ease data preparation. A major difficulty with such regular meshes is
unwanted proliferation of mesh nodes. For example, suppose some problem is investigated which involves a rectangular region, as in Fig. 11(a), in
which the area of major interest is in the lower left corner. To gain
sufficient accuracy and detail there, numerous mesh lines are densely

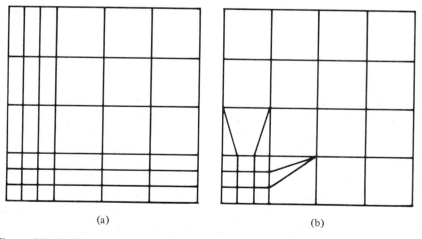

(a)   (b)

Figure 11. (a) Mesh in the *x-y* plane, with many finite elements in the area of
prime interest. (b) Similar mesh, but with detail outside the main interest area
suppressed by a singular transformation.

packed into that area. However, in a topologically regular mesh the lines cannot simply stop at the edge of the region of main interest; they must continue. The continuation of mesh lines thus causes many more finite elements and mesh nodes to exist in remote areas of little interest than are really wanted, as illustrated in Fig. 11(a). This problem is cured in some other systems by permitting singular coordinate transformations, which map two or more lines of the u-v plane into a single line of the problem plane. Such transformations lead to meshes as shown in Fig. 11(b).

The two meshes of Fig. 11 give the same amount of detail in the area of prime interest, but the number of mesh nodes, hence field variables, in Fig. 11(b) is only half as large as in Fig. 11(a). The difference clearly lies in transformation rules, which in Fig. 11(b) permit lines $u = 1$ to $u = 4$ to be coalesced into a single line in the x-y plane. Data preparation for such meshes naturally takes a good deal of extra care, for while adjoining mesh lines are allowed to coalesce into one, they are not permitted to *cross* each other!

Attribute Labelling

In contrast to MagNet, the data input arrangements of MAGGY are strongly oriented to the *mesh* of nodes and edges that bound the finite elements, but not to the elements themselves. In fact, elements as such do not appear in the mesh description at all. It is therefore not possible to identify material properties or sources in the mesh description, and separate blocks of statements are used for that purpose.

Material labels and source densities are attached to specific regions of the problem space, i.e., to specific sets of finite elements, by means of a *region block* of statements. A region block contains statements to identify materials and sources, and specifies the relevant geometric area by means of an *area statement*. For example,

$$AREA = X(1), Y(1), X(2), Y(2)$$

defines an area in the x-y plane which corresponds to a unit square in the u-v plane, by defining the contour that surrounds the area. In the u-v plane, this contour is shown in Fig. 12(a). It starts at (1,1) as specified by $X(1)$, $Y(1)$ and follows the latter line (the line $v = 1$) until it meets the line $u = 2$, then along that line until its intersection with $v = 2$. It is then tacitly understood to close back to (1,1) by following the mesh line $v = 2$ and finally the line $u = 1$. The location of the corresponding area in the x-y plane can only be determined by reference to the mesh statements which specify the coordinate transformation from u-v to x-y coordinates. A *region* is defined as one or more areas jointly containing the same materials and sources, so regions composed of several separate pieces, or multiply connected regions, can be specified by giving a set of areas. Each of these must be simply connected. For instance, the statements

```
REGION TWICE
AREA = X(1), Y(1), X(2), Y(2)
AREA = Y(3), X(2), Y(2), X(3)
BH = IRON
CURRENT = 1500.0
```

say that the region TWICE is composed of two simply connected areas, the *x*-*y* plane images of the two unit squares in *u*-*v* shown in Fig. 12(b). They are specified by the fourth statement to contain a material whose magnetization curve is identified by the name IRON. The final statement indicates that the total current through the two areas is 1500 amperes, with uniform density distribution as measured in the *x*-*y* plane.

Where complicated coordinate transformations are involved, area specifications can sometimes be tricky to get right. Furthermore, it is easy to make mistakes in specifying long strings of coordinates and coordinate lines, as is often necessary to describe complicated problems. The language employed to state the problem description is, when all is said and done, a programming language. A full problem statement thus constitutes a kind of program, just as prone to error as any other kind. It requires verification and debugging, and this is the sensitive spot where batch processing shows its disadvantages. On the positive side, batch programs acquire all data at once and can be arranged to test the data thoroughly for internal consistency before proceeding. MAGGY does in fact include a large variety of data checks, and allows most checks to be carried out without actually proceeding to problem solution.

Practical Mesh Construction

Using a good interactive mesh editing program, or a well designed mesh specification language for batch processing, the CAD system user is able to control precisely the relative placement of elements, the degree of

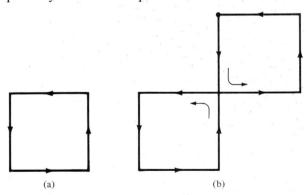

(a)   (b)

Figure 12. (a) A simply connected area in the *u*-*v* plane. (b) Two-piece region, described as two areas.

fineness in modelling detail, and the element shapes. However, control is useless without a good idea of what and why to control. How then should a mesh be chosen?

A few convenient rules of thumb for judging the goodness of a finite element mesh are set out in this section, the results of practical experience by numerous CAD system users. All such rules are empirical and of limited application, so they are often broken with impunity. At the same time, they all have some scientific basis, which it is well to keep in mind so as to be aware of the risks that might be run in breaking them.

Energy Minimality

Nearly all finite element methods, and many other numerical techniques as well, work by minimizing some quantity closely related to the stored field energy, which is locally a function of the flux density. The minimization is done by adjusting undetermined coefficients associated with the element functions: piecewise flat functions in first-order finite element programs, piecewise polynomials in programs using high-order finite element methods.

Finite element solutions will generally have very high accuracy if the shape of the exact solution is well approximated by the element functions. For example, if the true field just happens to have a piecewise constant flux density distribution, as for example inside a long uniform solenoid, then even first-order methods will yield exact solutions. Of course, flux distributions are rarely quite so uniform as that in practical problems. Nevertheless, the principle remains: a few large elements may be used where flux densities are nearly uniform, but numerous small elements will be needed where the flux lines curve sharply. In most meshes, a sound general rule (though one not always easy to obey) is to try to distribute equal amounts of error to all elements; in other words, to make the elements large where the error density is low, small where it is high.

A point to be noted carefully is that sharp curving of flux lines (i.e., rapid local changes in stored energy density) is of concern, not simply high flux densities. Large single elements will model the field well even in areas of very high flux density if the flux density is nearly uniform. Conversely, large or curious-shaped elements are just fine, provided the "grain" of the finite element mesh follows and favors the way the flux lines naturally wish to bend.

When working with axisymmetric meshes, the usual mathematical formulations deal not with the potentials themselves, but with the potential multiplied by a radially varying factor, most often r^{-1} or $r^{-1/2}$. In such cases, the above remarks still apply; however, what is extremized is the energy per radian, not the energy per unit length. In other words, near the axis of symmetry the triangulation may be somewhat coarser than would be the case for a comparable x-y plane problem, without ill effects.

The Equilateral Element Rule

Many experienced finite element analysts make some effort to observe a rule of thumb to the effect that *elements should generally be nearly equilateral*, and that if they cannot be made equilateral, they should in any case contain no included angles larger than about 170°, or what is equivalent, no angles smaller than about 5°. This rule is in a certain sense a corollary of the above discussion of energy minimality. Making all finite elements regular and equilateral makes the *geometric anisotropy* of the mesh, its "grain" in other words, more or less equal in all directions and thereby promises similar accuracy no matter which way the flux lines in the solution actually run.

Observance of the *nearly equilateral element rule* is only worth the trouble if the same mesh is to be used with an extremely wide range of different excitations so that radically different flux distributions may result, or if the user has no idea at all what shape the flux distribution may take in the solution (but can that user really be a designer?). Most practical designers therefore cheerfully but controlledly violate this rule.

The requirement that included angles be kept larger than about five or six degrees may be viewed as a corollary of the nearly-equilateral rule, since a restriction on included angles places an implied restriction on how far triangles can stray from being equilateral. However, the *five degree* rule has a scientific basis which goes beyond mere equilaterality. It results from the following considerations. When finite element equations are assembled, cotangents of the included angles are computed, either explicitly as trigonometric functions or, more frequently, implicitly through differences and quotients of side lengths. When angles smaller than about 5° (or larger than 175°) occur, cotangent values exceed 10, and one significant figure is lost to numeric roundoff. (By a similar argument, angles below about 0.6 degrees, whose cotangents exceed 100, result in loss of two figures.) In the algebraic process of solving matrix equations, finite element models of magnetics problems can be shown to lose about as much precision as the contrast of largest to smallest permeabilities in the problem. This ratio is typically 1000, so that the trailing three digits in well-converged solutions can be taken as meaningless. The usual 32-bit computer carries a little more than seven decimal digits in normal calculation. Allowing three digits to be lost to permeability contrast, and one to trigonometry (if angles are limited to 5 degrees), computed results may be expected to be good to about three significant figures.

Three significant figures in potential values is an acceptable, but not a very good, precision level in magnetic field calculations. It must be remembered that in most cases the analyst is primarily interested in flux densities and fields, only secondarily in potentials. But flux densities are obtained from potentials through a process of differentiation, which tends to enhance error. It follows that, with the 32-bit word length now common in computing, magnetic flux densities and field values are likely to be in error if very skinny, exaggeratedly long finite elements are employed. In the presence of ordinary magnetic materials, the limit lies at a minimum included angle of about 5 degrees, or equivalently, at a ratio of longest to shortest triangle side of about 10.

Like all rules, the five degree rule is often violated. Violations are likely to be successful in two sorts of circumstances. First, if elements of bad aspect ratio are placed in regions of low flux density, a local reduction in accuracy will take place, but it won't matter much. Secondly, in problems involving only one magnetic medium—for example, in solving for the field of an air-cored inductor—no significant figures are lost to material contrast, and some loss to small angles can be tolerated.

Symmetry in Meshes and Solutions

Geometric symmetry of the real device should be preserved in meshing whenever practicable. If asymmetry is introduced by the mesh, asymmetric solutions can be expected for physically symmetric problems. As an example, Fig. 13 shows the solution obtained for the no-load field of a turboalternator. This machine is geometrically symmetric about axes 45° apart, i.e., about the bottom and left edges of the drawing, as well as about a line inclined at 45° to both. In the solution shown, the excitations are symmetrically placed about the 45° line, antisymmetrically about the left and bottom edges. Consequently, the solution ought to show two identical lobes, but it does not. Its asymmetry is startling, especially to experienced machine design engineers who expect a perfectly symmetric field plot.

The reason for asymmetry in Fig. 13 is not far to seek. When Fig. 14 is examined, the finite element mesh underlying the solution is seen to be distinctly skewed. Although skewing may sometimes be desirable (e.g., in the interests of producing a minimal mesh), such biased meshes should generally be avoided.

Detailed examination of Fig. 13 will show that the results obtained with a skewed mesh are perhaps not quite so bad as might appear at first glance. The solution is very nearly symmetric in areas with high flux density, and therefore high stored energy density—the portions that contribute significantly to energy storage and energy conversion. Gross and striking asymmetry occurs primarily in areas where flux lines are far apart, i.e., where the flux density and stored energy do not amount to very much. Thus the solution of Fig. 13 is probably not far wrong mathematically. But the lopsided plot looks untidy, visually offensive, and interferes with whatever message the drawing may have been intended to convey.

Asymmetry is a difficult aesthetic problem to deal with. The human eye is extremely sensitive to regularity (or lack of it) in patterns, and observers readily spot even tiny deviations. Orthogonality and symmetry are two psychologically strong properties, and their absence where the eye expects them is readily apparent. In most cases, such gratuitous distractions

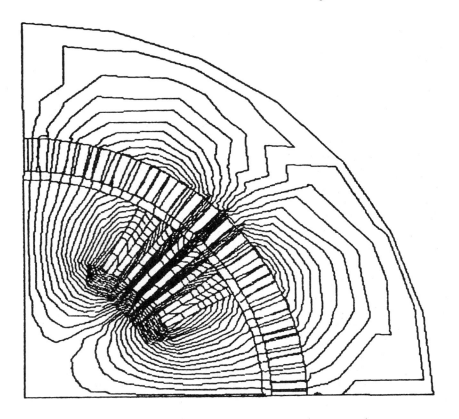

Figure 13. No-load flux distribution in an unloaded turboalternator, computed using an asymmetric mesh.

should be avoided by constructing meshes to have symmetry where appropriate.

Singularities and Reentrant Corners

The one really basic rule of mesh planning is to use a fine mesh where flux lines need to bend with a very short radius of curvature, a coarser mesh where their curvature is gentler and their spacing more or less uniform. This principle indicates that difficulty may be expected wherever reentrant corners are encountered, that is, at corners with interior angles in excess of 180°.

The meshing problems that result from reentrant corners are not always immediately apparent, as may be illustrated by a simple example. In Fig. 15, a segment of the interconductor space of a superconductive coaxial structure is shown. The structure comprises two square conductors placed

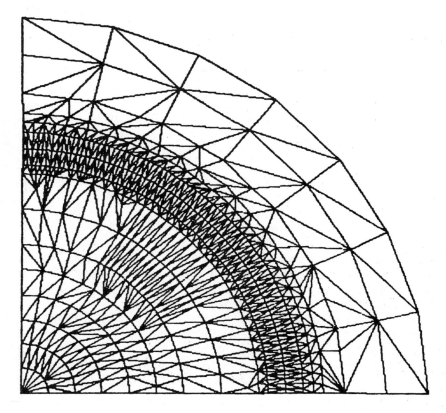

Figure 14. The mesh corresponding to Fig. 13. Asymmetry of the solution is entirely due to mesh asymmetry.

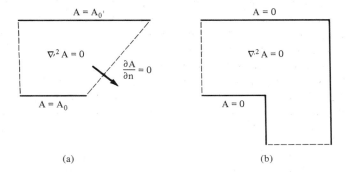

(a)   (b)

Figure 15. (a) One-eighth of a square caoxial structure, an apparently convex region. (b) One-quarter of the same structure, showing the "hidden" reentrant corner.

symmetrically with respect to each other. The inner conductor is assumed to carry current into the paper, the outer conductor serving as the return path. For reasons of symmetry, it is thus only necessary to analyze one-eighth of the device, as shown in Fig. 15(a). In the interconductor space, the magnetic vector potential must satisfy Laplace's equation. Since flux lines cannot penetrate into superconductive material, a flux line must follow the surface of the inner conductor, and another must follow the surface of the outer, so both conductor surfaces must be surfaces of constant A. The two symmetry planes shown by dotted lines in Fig. 15(a) must be crossed orthogonally by flux lines, if they are to be symmetry planes; they therefore form boundary conditions of the homogeneous Neumann type. So far as one can see from Fig. 15(a), the result is a well posed problem on a convex quadrilateral region, ready to discretize and solve.

That the problem of Fig. 15 in fact contains a hidden difficulty becomes clear when the drawing is enlarged so as to show explicitly both sides of the symmetry plane at the right edge of Fig. 15(a). Clearly, what appeared at first glance to be a convex region with an included angle of 135° in fact turns out to contain an interior corner of 270°, a reentrant corner. At such a corner the flux lines following the surface must bend with an infinitely small radius of curvature. The effect is to crowd flux away from the surface of the outer conductor near the bends, toward the inner conductor. At the very tip of the corner, the field will be singular.

The crowding of flux lines toward reentrant corners indicates that meshes ought to be constructed so as to use numerous small finite elements in their neighborhoods. Conversely, even a quite crude mesh will suffice near corners with an included angle less than 180°, as near the outer conductor surface in Fig. 15(b). An alternative, available in a few proprietary CAD systems, is to employ special elements which have the correct type of field singularity built into the element itself. This approach is elegant but it does demand an increased acquaintance with the underlying mathematical physics; it pleases analysts but often tends to confuse designers whose primary concentration is on device performance rather than field theory. But whichever kind of meshing may be employed in a particular system, the central message is the same: *beware the reentrant corner hidden under a symmetry plane.* Corner sharpness must not be judged in terms of a single geometric slice of the mesh, but in terms of the whole device, even if only a slice is to be analyzed by computer.

Clipping the Corners

Included angles greater than 180° lead to difficulties but, surprisingly enough, so can included angles below 180°. These difficulties are of a quite different kind, and arise particularly with first-order elements. They are again best appreciated by example.

A mesh, for some x-y plane problem in which the entire mesh perimeter is a flux line, appears in Fig. 16. A flux-line boundary in such a problem is a boundary of constant vector potential, say $A = 0$. Although the mesh may at first glance appear reasonable, closer examination reveals it to distort the actual region shape badly. If two edges of a first-order finite element are required to lie on the same flux line, then no other flux lines can exist in the same element; if there were any, they would either have to be curved (impossible in a first-order element) or they would have to cross other flux lines. To say the same thing in different words, the potential values in a first-order triangular element are the linear interpolate of the three vertex values, and if the three vertices are at the same potential, the entire triangle must have that potential. But the elements placed in the corners of the mesh of Fig. 16 all have two edges lying on the same bounding flux line, so these elements cannot contain any flux and are effectively excluded from the region modelled. In effect, the corners have been clipped off by the modelling process.

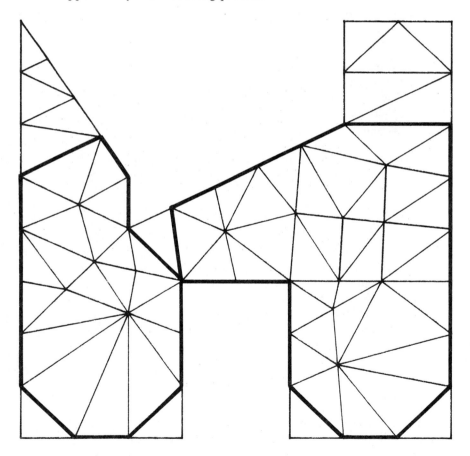

Figure 16. Mesh produced by an automatic mesh generator. The heavy outline shows the region actually modelled with first-order triangular elements.

A meshing phenomenon closely related to corner clipping, and even more worrisome, might be termed "bridging". In Fig. 16, the two reentrant corners near the narrow neck at the left share a triangular finite element, which has two of its vertices tied to the top edge of the neck, the third at its bottom surface. Since both surfaces are required to be parts of the same flux-line boundary, this element must have the same vector potential everywhere and therefore cannot contain any magnetic flux. It effectively serves as a flux barrier to cut the region modelled into two disjoint regions! For the same reason, four elements at the upper left, and three elements at the upper right of the drawing, must have constant potential values also. In consequence, the mesh of Fig. 16, which might appear quite reasonable at first glance, in actual fact cannot be considered satisfactory; all corners have been rounded, the two longish excrescences of the region have been truncated severely, and the region has been divided into two distinct pieces—all without intending to do so.

Exterior Boundaries

Many magnetic field problems concern devices surrounded by a field in unbounded space. To model such situations on a finite element mesh, the usual recipe is to bound the region with artificial boundaries sufficiently distant to leave the main part of the field unaffected. This class of problem is illustrated by the actuator solenoid field shown in Fig. 17. This device is discussed and its performance analyzed in detail elsewhere in this book; for present purposes, it suffices to note that the device comprises a solenoidal coil enclosing a ferromagnetic plunger, placed in air. Obviously, the air space needs to be modelled; but how much of it should be given a detailed mesh representation?

The practical question in modelling is just how far away the artificial boundary should be placed. Clearly, placing it close by means very small expenditure in modelling effort and low computing cost; placing it far away, on the other hand, implies higher accuracy. To gain some idea of how far is far enough, the following simple argument will be useful.

Consider two equal and opposite point magnetic poles of unit strength, as in Fig. 18. The magnetic scalar potential at the point P, due to the two poles, is given by

$$\Omega = \frac{1}{4\pi} \left[\frac{1}{r_1^2} - \frac{1}{r_2^2} \right]. \tag{4}$$

The two distances r_1 and r_2 may be written in terms of the dipole spacing d and the distance r from the observation point P:

$$r_1^2 = r^2 + (d/2)^2 - rd \cos \theta \tag{5}$$

and

$$r_2^2 = r^2 + (d/2)^2 + rd \cos \theta \tag{6}$$

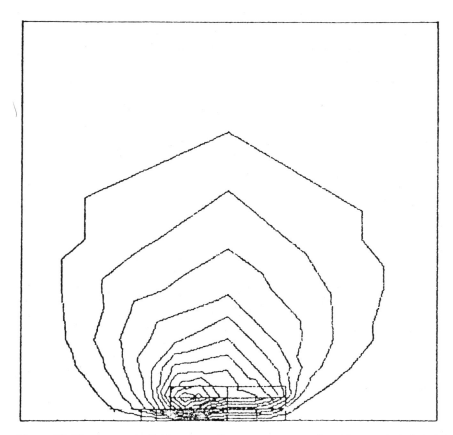

Figure 17. The magnetic field surrounding an actuator solenoid extends infinitely far but may be truncated at a convenient distance.

If the distance r from the observation point is large compared to the dipole spacing d, $r >> d$, then the potential expression (4) is well approximated by

$$\Omega = \frac{1}{4\pi} \frac{2d \cos \theta}{r^3}. \tag{7}$$

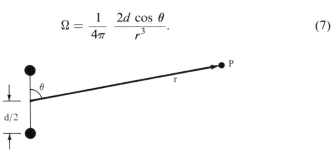

Figure 18. The potential of a dipole can be expressed in terms of distances d and r.

In other words, the potential falls off as the cube of the distance r from a dipole. This very rapid decrease permits the artificial boundaries to be placed rather close by the region of interest in many problems. Most magnetic devices of course are not dipoles, but something more complicated. Their potential fields have Taylor series of the form

$$\Omega = \frac{1}{4\pi} \left[\frac{a_3}{r^3} + \frac{a_5}{r^5} + \cdots \right], \tag{8}$$

whose leading term has the same behavior with r as the approximate expression of equation (7). Since the higher-order terms fall off even more quickly with increasing distance, they may be safely neglected; in other words, *viewed from a distance of two or three diameters, any object resembles a dipole.* As a general rule of thumb, most problems can therefore be solved adequately by placing an artificial boundary a few device diameters away. If high-precision results are required, equation (8) may be used to estimate the distance r at which all but a prescribed percentage of the total stored energy will be included in the problem; since finite element methods are generally energy-minimization methods, that estimating basis is a good one.

The above argument holds for three-dimensional problems, and for two-dimensional problems in r-z coordinates. A similar argument can be made for two-dimensional translationally symmetric problems (problems expressed in x-y coordinates). In that case, the approximating series contain a term in r^{-1} as well, and the potential drops off less rapidly than in the three-dimensional case. Estimates for reasonable placement of artificial boundaries can be similarly arrived at, however.

The quantity examined above is a potential, not the magnetic field. The field $\mathbf{H}$ is determined by the gradient of potential and therefore falls off more quickly than the potential itself, behaving as r^4 and r^2 in the r-z and x-y cases. The energy densities, which are the quantities computed in the finite element method, depend on the square of field and therefore fall off as r^8 and r^4 respectively. It may be concluded that even if a mesh needs to be extended out to a distance of several device diameters, a quite rough mesh will suffice everywhere except very close to the region of prime interest.

Skin Depth in Diffusion Problems

In solving eddy current problems and other problems involving diffusion phenomena, fields in the solid bulk portions of diffusive objects are often small. It often suffices to model in detail only the portions near the surface where currents are actually expected to flow.

An estimate of how far, and in how much detail, to model may be gained by considering that in a planar slab of conductor, eddy currents fall off exponentially with depth, reaching $1/e$ of their surface density in a distance equal to one *skin depth*. Such behavior is usually adequately modelled by two or three first-order finite elements, or one element of second or third order, per skin depth. Reasonable models can thus be obtained by estimating the skin depth and providing the appropriate fineness of modelling near the surface. Of course, there is little point in modelling in any detail at all beyond a depth of three skin depths or so, for the field has decreased to less than 5% of its surface value at a depth of three skin depths and keeps dropping thereafter.

Many eddy current and skin effect problems will involve geometric complications as well as diffusion phenomena. All comments made above regarding reentrant corners, artificial boundaries, and element shapes, will still apply, *mutatis mutandis*.

Automating the Mesh-Building Task

The creation of mesh models frequently involves repetitive operations, work which is generally best left to machines because people have rather less patience and make more mistakes at it. Two approaches are commonly employed for automating the task: automatic mesh generation and editor programming. The two are not mutually exclusive; there do exist CAD systems which contain at least elements of both.

Approaches to Mesh Construction

Two approaches for defining geometric data are commonly followed by CAD systems. They may be termed *synthetic* and *analytic* respectively, though the difference is often one of viewpoint more than technology; indeed the two blend almost imperceptibly in many real systems.

The synthetic approach to meshing regards the geometric shape as being the union of a set of finite portions or regions, so that the geometric shape is implied by the mesh of regions. The regions may variously be finite elements, or composites made up of finite elements. In the latter case, the region shapes are often stored, along with prescriptive parameters to indicate how the region is subdivided into elements, but actual subdivision into elements is done only when the analysis is performed. In other words, the subdivision algorithm is specified for each region and executed at some time by a simple automatic mesh generation program. If carried to a high level of sophistication, this approach eventually becomes indistinguishable from the analytic approach.

In the analytic approach to mesh building, geometric shapes are explicitly encoded and stored, using one of the available computational geometry systems. Geometric modelling systems know nothing about finite elements and store object descriptions purely in terms of geometric primitives—spheres, cylinders, blocks, and other basic shapes, as well as unions and intersections of these. The finite element mesh with which the analysis is to be performed is generated subsequently by an automatic mesh generation program. All user interactivity is concentrated in the geometric modelling portion of the work; the mesh generation is a straightforward batch program run even if it is embedded in an otherwise interactive system.

Fully automatic mesh generation is as yet rare in practical CAD systems, for only very recently have entirely reliable computational algorithms been developed and published. Where manual editing of the mesh is possible, occasional failure of the automatic program is a minor blemish; but where manual alterations cannot be made, meshes containing geometric or topological errors (e.g., overlapping elements) become serious problems even if they occur only rarely. As a result, completely automatic mesh generators are very rare, while limited forms of automatic mesh generation or mesh refinement are employed by virtually every current program, as aids to manual work or as data compression devices. For example, in the PE2D system quadrilateral regions are used to define subsequent automatic mesh refinement operations, by attaching to them further information regarding the desired refinement scheme. In the MagNet system, minimal triangulations of essentially arbitrary polygons are generated by the MAKE command, permitting subsequent refinement. In other words, the analytic approach to mesh generation is rarely seen; the synthetic approach, with analytic assists, is common.

In general, automatic mesh refinement procedures place severe demands on computing resources, but require little if any user intervention. They are therefore well suited to batch processing, and also to programs designed to be used from remote terminals with poor interactive response. The synthetic approach, in which regions are built up from elementary parts, is better suited to working environments with good interactivity.

Mesh Synthesis

In the synthesis approach to geometric description, the problem region is regarded as being built up of standardized types of subregions—for example, triangles. This viewpoint is obviously well suited to finite element analysis if every region also coincides with a finite element used in eventual analysis. However, it is not necessary for the various subregions to be finite elements; they may be unions of simpler subregions, each of which is in turn a union of subregions, and so on. For example, the composite region $\{A,B,C,D,E,F,G\}$ shown in Fig. 19 may be regarded as the union of two subregions,

$$\{A,B,C,D,E,F,G\} = \{A,D,E,F\} + \{B,C,G\}, \tag{9}$$

which in turn are unions of subregions,

$$\{A,D,E,F\} = \{A\} + \{D,E,F\}, \tag{10}$$

$$\{B,C,G\} = \{B,C\} + \{G\}. \tag{11}$$

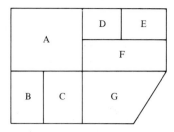

Figure 19. A two-dimensional geometric region subdivided into subregions, which might represent finite elements.

Two of these may in their turn be regarded as further unions of subregions,

$$\{D,E,F\} = \{D,E\} + \{F\} \qquad (12)$$

and

$$\{B,C\} = \{B\} + \{C\}. \qquad (13)$$

Finally,

$$\{D,E\} = \{D\} + \{E\}. \qquad (14)$$

The entire geometric problem model of Fig. 19 may therefore be described by the hierarchical structure of Fig. 20, in which the rightmost components are finite elements, while the leftmost level of the hierarchy represents the entire mesh. In this view, the synthetic approach to mesh construction may be regarded as proceeding leftward from the elements, while fully automatic mesh generation proceeds to the right. Intermediate stages are of course possible, in which composite regions, such as $\{D,E,F\}$, are specified by the user. Indeed almost all currently existing mesh generation programs work at least partially in both directions.

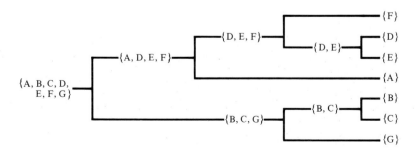

Figure 20. A hierarchical data structure (tree structure) representation of the geometric model of Fig. 19.

Programming of Editing Operations

Speedy construction of complicated models is obviously best accomplished by a two-step process: planning and building. The planning step consists of devising a decomposition of the structure into parts, the parts into subparts, so as to reduce the need for detailed modelling to handling only a small number of elementary sections or parts. In terms of Fig. 20, the object is to choose a decomposition which makes as many leaves of the tree structure, and as many branches as possible, either identical or very similar to each other. This step is usually carried out entirely by hand, without any computational aids. Building, the second step, is the reverse of planning, but carried out with the help of the computer: every elementary part is built, then edited and joined to others, continuing until the whole has been constructed.

The initial decomposition step in model building cannot easily be relegated to a machine, for only the designer knows what he really intends to build. But the synthesis of a model out of elementary bits is clearly a proper task for the computer. In fact, the synthesis procedure can be mechanized to a very large extent. It is only necessary to allow editing commands to be accumulated into *programs* or *command files*, which are executed at a later time, on request. For example, the MagMesh mesh-building system permits such deferral of commands, thereby becoming a *programmable* system. In other words, the commands in the MagMesh language can be used to write stored programs; they do not need to be executed on the spot.

To defer execution of a command in MagMesh, i.e., to place it in a program file instead of executing it immediately, the command itself is preceded by a line number. For example, the command conversation

```
>> MACH ready.  Command:  20 COPY TEMP KEEP
>> MACH ready.  Command:  40 MIRROR
>> MACH ready.  Command:  70 JOIN TEMP
>> MACH ready.  Command:  90 DELETE TEMP
>> MACH ready.  Command:  LIST
```

enters a set of commands, then lists the stored program on the terminal screen, as

```
20 COPY TEMP KEEP
40 MIRR
70 JOIN TEMP
90 DELE TEMP
```

LIST is one of several commands that manage and control stored programs.

The JOIN command of MagMesh, or equivalent *union* commands in other programmable editing systems, may be used to implement the tree structure of Fig. 20. Thus, Fig. 20 directly leads to an algorithm for constructing a mesh, hence to a program in the MagMesh editing language. The main difference between the data structure of Fig. 20 and the program is, aside from notation, the fact that the data structure is mathematical and static—it merely expresses certain truths—whereas the program is algorithmic and dynamic.

The data structure diagram of Fig. 20 is easily turned into a programming flow diagram, by merely reversing the direction of its flow, as in Fig. 21! The flow diagram can then be converted directly into a program, as shown in the right half of Fig. 21. To store the program and make it a permanent part of the user's repertoire, each command is prefixed with a line number.

Once a program is created and stored, it may be re-used at any time. In fact, it may be made into the definition of a new command verb, thereby enabling the user to customize the system to his own personal needs. The use of stored programs makes it very easy to create variations on a design or to investigate the effects of local shape alterations. To create a variant design, it is frequently only necessary to change one or a few of the elementary part models, then to allow the large model to be reassembled under program control without further user intervention.

Automatic Mesh Generation

Some CAD system designers, and some system users, believe that the finite element mesh on which the solution is computed should not concern the user at all. They feel that the finite element mesh is an intermediate artifact of the analysis method; the user should be interested in the shape of the object analyzed, not the shape of the mesh. In this view, the user should build a geometric description of the device to be analyzed, then allow a fully automatic mesh generator to construct the mesh.

Fully automatic mesh generation is an aesthetically pleasing goal to aim at, but for the moment a very difficult one to achieve. There do now exist many sophisticated geometric modelling programs, which allow

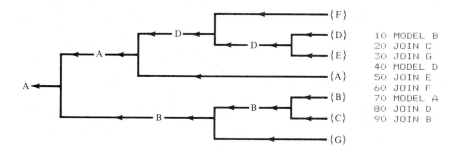

Figure 21. The model of Fig. 20, restated as a mesh-building algorithm and accompanied by a program segment in the MagMesh editing language.

computer-readable descriptions of geometric shapes to be built up. However, many of the problems which occur in magnetics involve very complicated shapes, which still must be communicated to the computer. Thus, automating the mesh generation may remove some chores from the user, but may not reduce the work to be done by very much. Specifying the part shapes for the shaded-pole motor of Fig. 22, for example, is a task almost as lengthy and demanding as the construction of its finite element mesh. The automatic generation of meshes for such situations is desirable for many reasons, but it is not likely to be a major labor-saver.

For the long-term future, automatic mesh generation undoubtedly shows great promise. Experience to date indicates that at least partially automatic mesh generation is essential in three-dimensional problems, which are very much more demanding than their two-dimensional counterparts, especially of the user's ability to visualize and describe solid objects. Fully automatic finite element mesh generation from geometric modeller output is possible in principle; but the element types for which a complete theory exists are severely restricted. No commercially available

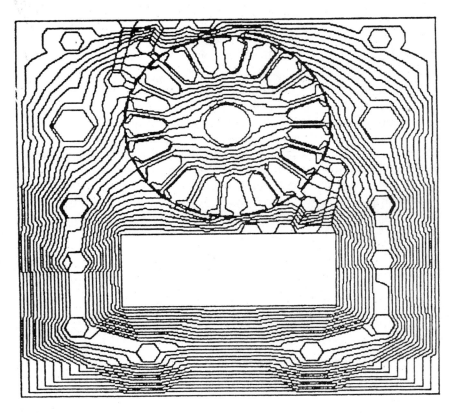

Figure 22. A small shaded-pole motor, analyzed using a mesh of about 1000 triangular elements.

packages for electromagnetics CAD perform entirely reliable automatic mesh generation from geometric modeller output, though some proprietary packages of major industrial laboratories have been known to offer largely automatic mesh generation for several years. These generally use ad hoc methods which work most of the time, but occasionally malfunction; they are perfectly acceptable in research and development environments where users are highly knowledgeable, and where the programs are maintained and serviced by their original designers on the spot. Commercial-quality packages, able to survive in the less forgiving atmosphere of the design office, will likely appear on the market in the late 1980s.

An objection to fully automatic mesh generation is sometimes raised by sophisticated users, who feel that automatic methods produce meshes not well optimized to any one purpose. Just what constitutes error in an analysis is partly a subjective matter: the user interested in terminal impedances may not much care whether local field values are accurate too. Since an automatic mesh generator does not know what aspect of a problem the user is interested in, it cannot produce optimal meshing. Hence some users are quite keen to retain manual control of mesh generation, just as some drivers are insistent on a manual gearbox in preference to an automatic transmission.

One very interesting, and very promising, future development is *adaptive meshing*. The principle is simple. To start, the problem to be analyzed is described by a very crude finite element mesh, and a rough solution is produced. Error estimates are made for this solution (they do not need to be very accurate) and a new, more refined, mesh is produced. A solution is computed on the finer mesh, an error estimate is made, then another solution, and so on until the solution has the desired accuracy. Adaptive methods are also at the prototype stage as of the time of this writing; they should be appearing in the marketplace in the next five or ten years.

Assembly of Problems

Definition of the geometric shapes and finite element meshes to be used in analysis is not sufficient in itself, because the abstract labels which identify elements and constraints are still undefined. The correlation between element labels and materials must be stated, and boundary values as well as source densities must be specified before a complete, solvable problem results. These tasks are usually referred to as *problem definition* or *problem assembly*. The tasks of mesh building and problem assembly are quite distinct; if mesh construction may be viewed as computational simulation of building an experimental prototype, then problem assembly must be regarded as analogous to devising and setting up an experiment to test the prototype. The work of problem assembly is usually handled by entirely separate program code. This detail of CAD system construction may not be visible in batch-run systems such as MAGGY, where all data must be furnished to the computer at the same time, but it becomes obvious in interactive systems such as MagNet, because an entirely separate system module is used.

Representation of Physical Problems

A problem fully defined and ready for computation must contain three ingredients: (1) a geometric model, (2) a file of the relevant material properties, and (3) a problem data table. The geometric model includes the finite element mesh to be used; it has already been discussed at some length. If computer analysis is to be viewed as simulation of prototype construction and testing, the analogous elements would be (1) a description of the prototype device, giving its size and shape, (2) properties of the materials of which it is made, and (3) a description of the experiment to be carried out. Obviously, many distinct problems can be associated with a particular geometric model, just as many experiments can be carried out on a given prototype.

In the MagNet system, a *problem* is defined as a data structure which is associated with a given geometric model. It is referred to by its distinctive *problem name*, and it contains the following ingredients:

1. A descriptive text or title to identify the problem.
2. A table of values corresponding to all constraint labels.
3. Definitions of source densities for all element labels.
4. Identification of the material corresponding to each element label which occurs in the model.

It should be noted that the data structure called a *problem* consists only of material and source identifications and contains no geometric or topological information. A physical problem must therefore always be identified by two names: the name of a model and the name of a problem defined on that model. As indicated in Fig. 23, both names must be specified.

Problems are contained in data tables which are stored in files. They are therefore subject to all the usual considerations that apply to files. That is, there must be facilities to edit any given problem, as well as to delete it from the files, to store it away, to retrieve it again, and so on. Problem definition itself is thus another editing process, one that establishes or alters the data tables or lists in which the problem is stored. The exact form of data tables is naturally very much system-dependent; the principle, on the other hand, is broadly applicable.

The problem data tables as used in MagNet are made up of two principal segments: a table of constraint label meanings, and a table of element label meanings. The constraint definition table refers to binary constraints in the specific form

$$A_i + k_{label} A_j = R_{label}. \tag{15}$$

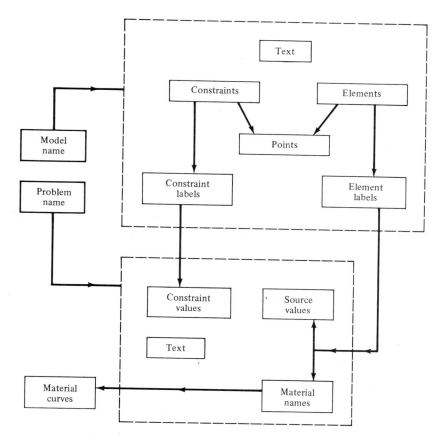

Figure 23. Relationship of data entities in a problem. The upper dotted line encloses the *model*, the lower one the *problem* definition.

It lists the multiplier k_{label} and the right-hand side R_{label} corresponding to each label that occurs in the model. An example of such a table is shown in Fig. 24. The element definition table is similar in principle: it lists the source density and the material name corresponding to each element label occurring in the model. An example of such a table appears in Fig. 24 also. Since only the material names, not details of material properties themselves, are given in this table, the problem is not fully defined until a material curve library exists, in which all the necessary materials are defined in detail. In other words, the element label table does not define the materials, it only gives references to another data entity, the materials library.

Time-harmonic phenomena are not catered for in the prob'em representation given above; neither are axisymmetric problems, nor permanent magnetic materials, nor indeed a great many other cases of interest to the magnetic device designer. All such are accommodated by extending the problem description tables. For example, time-harmonic

Element Labels

La-bel	Mate-rial	Source density
A	AIR	0.000E + 00
I	IRON	0.000E + 00
W	AIR	5.000E + 05
X	AIR	5.000E + 05

Constraint Labels

La-bel	Con-stant	Right side
H	0.000D + 00	2.500E + 00
N	0.000E + 00	0.000E + 00
P	−1.000E + 00	0.000E + 00

Figure 24. Two tables suffice to define problems completely: one for element and one for constraint labels.

eddy current problems require a value of conductivity to be associated with each material, not only magnetic properties; furthermore, the excitation frequency must be specified.

Problems periodic in space need to have the periodicity specified, as a global numeric parameter somewhat similar to frequency in time-dependent cases. However, there is a significant difference between axi-periodic problems to be solved in the r-z plane, and longitudinally (z axis) periodic ones requiring x-y plane solutions. In the former case, the problem is periodic in the angular variable θ, which is always bounded in the finite range $0° < \theta < 360°$; only full cycles can occur in this span of angles, and the periodicity is best described by stating how many full cycles occur in the full circle. This number is of course always an integer. In contrast, phenomena periodic in the longitudinal variable z may have any period whatever, because z is not bounded. The best way of specifying periodicity is thus to give the length of a period.

Problem Editing

Completely defined problems ready for solution are, from a computing viewpoint, data files ready to be communicated to a solver program. Assembling problems for solution thus means creating these data files, an editing process best carried out by a suitable editor program.

Establishing the label meanings in a batch-run system such as MAGGY is straightforward: the data tables are set up by stating their contents in appropriate input statements. The commands themselves are strings of words and numbers, so that the editing required is best carried out by a text editor similar to that used for preparing any other alphanumeric text, from Fortran programs to administrative correspondence. The only difficulty encountered is the lack of graphic communication, which makes blunders relatively difficult to spot.

In an interactive system such as MagNet, problem setup is carried out using a graphics-based editing program specialized to problem files. Mag-Prob, the problem editing module of MagNet, initiates a problem by assigning it a name and giving values to all problem variables. A fully defined and valid problem is therewith defined, although its variable values are probably not those the user desired. In the absence of contrary information, MagProb assigns default values to all variables: all materials are taken to be AIR, all source densities are set to zero, all unary (e.g., Dirichlet boundary) constraints are made to have zero value, and all binary constraints are made purely periodic (multiplier $= -1$, right side zero). The user may then alter this valid but uninteresting problem by editing it. The principle is thus similar to geometric mesh creation: a data set is created immediately, valid but probably unwanted; it is turned into the desired one by editing.

Problem editing in MagNet proceeds on much the same basis as editing of meshes or material curves. The tables associated with a particular problem are displayed, and the user is asked which values to alter. Alterations are made by a combination of graphic and keyboard actions. For example, to specify the total current in a winding as 120.7 amperes, the user points to the graphic display and identifies the winding by means of a cursor hit. The system responds by highlighting all sections of the model containing that particular material label. This graphic echo serves both to verify which part is meant and to remind the user (where appropriate) that the region in question may consist of multiple disconnected pieces. Selection (i.e., identification of the winding) is thus carried out entirely by graphic means. The current value, on the other hand, is entered through the keyboard, as

`120.7 TOTAL`

where the word TOTAL indicates that the numeric value given specifies the total current, not the current density. Similarly, materials are identified by pointing at the relevant part of the model and entering the material name at the keyboard.

Constraint editing follows a similar procedure. Identification is made by pointing at a constrained point. The system response consists of highlighting all points with that particular constraint label in the drawing. In the case of binary constraints, lines are drawn between corresponding point pairs. Again, numeric data must be entered by way of the keyboard.

In addition to the editing commands there are also, as one might expect, a few commands for deleting and renaming problems, for listing the problems currently on file, for recalling a problem to permit more editing, and for housekeeping functions such as shutting down the system.

If MagProb is employed to set up a sequence of problems, as is frequently the case in design studies, each newly defined problem is initiated by making an identical copy of the preceding one. This arrangement implies that where only one or a few parameters change from problem to problem (e.g., only the current density in one winding), editing operations are required only to specify the *changes* from one problem to the next, not to set up anew all parameters for each problem.

Solver Directives

In most cases, the solution of finite element equations requires a substantial amount of processor effort and very little user interactivity, if indeed any. This situation contrasts sharply with the various editing and postprocessing operations, whose central processor requirements are generally quite modest, but which place high demands on input–output subsystems. Most CAD systems therefore set up their equation-solving procedures as what really amounts to batch jobs.

How equation-solving jobs are actually set up varies widely from system to system, even though the overall performance characteristics are similar. In some cases, for example in the FLUX2D system, users are actually required to submit them as separate tasks to the operating system. In other systems, submission is handled automatically but still performed in a straightforward run without user intervention. The details of automatic handling depend on the computer system design, both as to the hardware available and the operating system in use. Where a single time-shared central processor is in use, solution jobs may be queued as separate low-priority tasks, which remain inaccessible until they have run to completion. Where several central processors are available in the computer system, as in some implementations of PE2D, separate submission to another processor may not be necessary; but since the solution task is handled by another machine and another operating system, no significant intervention with this process is possible until it has completed its run. In every case, equation solving acquires the "ballistic execution" aspect of a batch job: all user decisions must be made beforehand, for no intervention is possible after launch, other than perhaps aborting the mission.

Because equation solving is a process with limited interactivity, the user must generally select in advance all the options and choices offered by the solver program. In the MAGGY system, for example, two different mathematical methods of equation solving are available. The user must specify which one is to be used, and must also state the expected degree of precision. One of the methods is semi-iterative, and the user may prescribe the permissible number of iterations per solution cycle, in addition to deciding on convergence criteria. In a similar fashion, the solver programs of MagNet allow the user to steer ("to aim" might perhaps be a more apt verb) the solution process. The choices available in MagNet include two error tolerances, since a modified Newton technique is used for the nonlinear problem and iterative solvers are employed for the linear problems that occur at each Newton step. The modified Newton scheme involves several further parameters which deal with the Newton startup procedure, the maximum number of Newton steps to be allowed, and the like.

The quite large number of choices open to the user are communicated through *solver directives*, instructions to the solver on how to behave and what to do. Setting up all solver directives is considered a job for MagProb in the MagNet system. Since the solver directives must take the structural form of another data table, another editing facility is required somewhere along the line. Rather than employ yet another editor, MagNet combines the editing of solution parameters with the editing of problem data tables; the solver directives thereby effectively become part of the problem definition. A substantially similar approach is used by PE2D. Batch-run programs like MAGGY handle solver directives in much the same way as any other data, by providing a suitable set of commands that can be read from an input stream. MAGGY actually makes the solver directives into a separate block of command statements, the *algorithm block*, which occupies a role analogous to the region block, the curve block, the print block, etc.

Default Values

Choosing solver convergence criteria, limiting iteration numbers, and setting starting parameters are mysterious activities to many CAD system users, and help must be provided by the system. Solvers, after all, form a highly mathematical and quite complicated part of any CAD system; yet their internal workings are not generally obvious to the user, nor even of particular concern. To most users, they are somewhat like the engine in a car: their performance matters, but the details of compression ratios and cylinder displacements are of minor interest at best. Setting up solver directives is therefore regarded as a specialist job by many, something to be avoided if not absolutely essential.

The most common, and probably the best, way of helping the user is to have the system itself choose reasonable parameter values by default. While it is impossible to do justice to all possible cases and circumstances through clever default choices, most CAD systems do set up solution procedures so that a user not wishing to delve into the underlying mathematics need not bother. In MagNet, for example, the user must explicitly wish to edit the current solver parameter settings. Similarly, MAGGY assumes default values if an algorithm block does not specifically declare them. On the other hand, users already initiated into the mysteries of preconditioned conjugate gradient methods may well wish to choose their own tolerance levels or convergence parameters and are entitled to do so.

Unlike most other current systems, MagNet does not distinguish between "solved" and "as yet unsolved" problems; unsolved problems are viewed as being similar to solved ones, but with a very large probable error in the solution. Whenever a new problem is created in MagNet, it is automatically assigned a file area of the size required to house its solution, and the solution is accompanied by an error estimate. Initially, all unknown potential values in the file area are set to zeros, and the estimated error in this default solution is taken to be 100%. Formally, the problem is thus considered to be accompanied by its solution, albeit one which is rather inaccurate until it has been improved by a solver program. Indeed, from this point on MagNet considers any and all recourse to solver programs to constitute *solution refinement*. In this way, "unsolved" problems never exist, only problems whose accompanying solutions have a large error.

Iterative solver programs can be used to refine already existing solutions if the process of solving does not alter any file formats. Although solvers usually run as batch programs, the principle of file identity is similar to that described for editor programs: if the input and output files of a process are identical in form, then second thoughts and changes of mind become possible without having to restart the entire job. For example, it is perfectly feasible to request that a particular problem be solved to an accuracy of 2%, then to undergo a change of heart and ask for further refinement to 0.5% instead, without restarting. It is only necessary to use the solver output file as its input file in another pass; so far as file structure is concerned, both are the same thing, therefore accessible to the same programs.

Field Problems in Magnetic Devices

Field problems as handled by the analyst are mathematical or computational abstractions of reality. The real physical world's rich complexity of detail precludes taking account of any but a few selected features of a device in the process of analysis, so it becomes the responsibility of the design engineer to decide what aspects to include, which to ignore, and which to represent in an approximate fashion. Although there are many good computer programs to perform every mathematical step of the analysis itself, no computer can undertake to select which features of a device are important, and which results are desirable. The art of magnetic design in the brave new world of CAD bears some resemblance to the craft of the graphic artist in a world endowed with photography. Compared to an earlier age, there is less need for detailed mastery of the trade secrets and craft tricks of algebraic manipulation and calculation—the machines can handle much of that—and an ever increasing need for sensitive perception in selection, abstraction, and interpretation. Accordingly, the formulation of problems, rather than their solution, comes to occupy a primary role, as this chapter seeks to illustrate.

Analysis in Design

Electromagnetic devices generally are built to satisfy specified performance criteria. Some of these criteria will be laid down explicitly, to reflect the wishes of the customer as well as other external influences (e.g., safety legislation). Others may be imposed by intrinsic considerations, such as the properties of materials and manufacturing processes. The designer's job is to ensure that the customer's specifications are met within the bounds established by materials, processes, and costs. Whether they really are, prototype trials ultimately decide. Whether they probably will be, on the other hand, can be forecast with a high degree of confidence through detailed analysis. Analysis in design really amounts to a simulation of prototype construction and test, and computational analysis serves above all to ensure that only one prototype need be built.

Explicit and Implicit Specifications

Design criteria stated by the ultimate user of a device often appear in the form of bounds on global quantities such as equivalent circuit parameter values or overall efficiencies. Intrinsic criteria are more often of a local nature; for example, the maximum flux density permitted in a device may be subject to an upper bound. The global, explicitly stated, kind of quantity can often be evaluated on a more or less automatic basis, indeed special-purpose programs are sometimes written to accomplish the job. The latter, implicit, specification is often less precise in fact even if it seems to be stated with precision. To continue the above example, the designer may be more confident of success if flux densities everywhere are comfortably lower, but not a very great deal lower, than the maximum permissible value; he will know that he is making effective use of material while still maintaining some margin for error. An automatic program that merely reports the maximum value of flux density and its location in the device only tells a small part of the story. How close to the maximum does the rest of the material come? Where is the next highest value? Is a small variation of shape or material properties likely to affect the result significantly? Does the maximum value occur in an area where simplifying assumptions might upset accuracy particularly badly? Are there any places where the flux densities are particularly low, where material could be removed?

To answer the various questions that interest the designer, analysis and design need to interlock; indeed,they have traditionally done so. Because magnetic devices are complicated, analysis is usually based on physical assumptions and approximations suggested by earlier designs. New designs in their turn usually rely quite heavily on analysis. Not only is it cheaper to analyze than to build and test; analysis often yields information which no reasonable tests can possibly furnish.

Bounding Analysis Complexity

To answer the real questions often asked by designers, and to answer them reliably, only a full three-dimensional analysis with hysteresis and material anisotropy taken into account will be certain to suffice. Unfortunately, such analyses are not generally possible; and even where they can be reasonably attempted, their cost is exorbitantly high. Not only do they cost computing time (which becomes cheaper every day), but they have a gluttonous appetite for man-hours. To read, interpret, and act on the vast quantities of data that make up a full three-dimensional solution takes a great deal of experience and consumes prodigious quantities of time. Practical analysis therefore must be based on simplified models, and the analyst's craft is one part computation, three parts skill in problem abstraction and formulation.

It is likely that the best currently possible analysis will always be expensive, for although the cost of any particular computational method decreases from year to year, that very decrease itself encourages the development of ever more powerful and costly new methods. But the full cost of a detailed analysis should be paid for only one or a few final designs, preparatory to the construction of prototypes; most candidate designs should be eliminated from consideration by simpler methods of analysis just as detailed analysis eliminates them from the prototype stage. In an ideal world, the designer should have at hand a series of analytic tools of ascending power and complexity, with (presumably) correspondingly ascending costs, so that designs can be refined stepwise, using analysis methods of increasing sophistication while moving closer and closer to the final design. There are two closely related general techniques for keeping the complexity of successive analyses within bounds: *successive refinement* and *subproblem analysis*.

The major tool for bounding complexity (which in the final analysis means cost) is *successive refinement*. Its object is to alter both the mathematical method used and the physical models employed, so that in any given design study the number of cases to be analyzed diminishes at least as rapidly as the cost per case rises. The varieties of physical models commonly available for this purpose include (1) magnetic circuit analysis, (2) two-dimensional field analysis, (3) augmented two-dimensional analysis, called *two and a half dimensional* by some, and (4) full three-dimensional modelling. Within each one, a wide range of variation is still possible when it comes to fineness of mesh construction and the degree to which detail is included or suppressed. Successive stages of analysis may therefore pass through several levels of modelling complexity on their way to the prototype.

Magnetic circuit analysis is in essence a technique for one-dimensional magnetic field solution, feasible with pencil and paper, and widely known to design engineers. Two-dimensional analysis is now routinely available in commercial software systems, for a range of problem types. Two-and-a-half-dimensional analysis is almost as easily available, but less easy to use because the interpretation of results, as well as the formulation of problems in the first place, is not quite so easy. In other words, the difficulty tends to be one of user expertise, not software capabilities. The three-dimensional packages which are gradually becoming commercially available are suitable for very simple problems, but the art of formulating and interpreting three-dimensional problems is still near enough the research forefront to require substantial effort even by knowledgeable users.

The second basic approach to bounding the complexity of analysis is *subproblem analysis*. In this technique, detailed portions of a problem are replaced by simpler ones, which are likely to be quite wrong internally but are expected to yield the right external result. The simplified partial model is obtained by refined analysis of only that part of the problem. This approach is highly traditional; it appears in a formal way in electric circuit theory as the idea of a *terminal equivalent*, and it is almost second nature to most designers. An early example of subproblem analysis may be found in the work of Carter, who established that it was not necessary to model the very intricate geometry of electric machine teeth when attempting to determine terminal parameters; it sufficed to replace the teeth and slots by a geometrically much simpler part whose permeability was neither that of air nor that of the machine iron, but an intermediate value given by the *Carter coefficient* well known to electric machine engineers. The aggregation of complex details into externally equivalent, but geometrically simpler, parts is usually achieved by detailed analysis of subproblems, ignoring the world external to the subproblem altogether.

Analysis by Successive Refinement

The process of successive refinement usually begins with the simplest analytic method available and progresses to increasingly complex models. Successive refinement is not inherently a hierarchical process, and every stage of refinement need not include within it all the detail of its predecessor. Not infrequently, it is tactically wise to fan out refinement in several directions, constructing several models of moderate complexity with different detailed features emphasized in each one. This approach to modelling is applied to a fairly simple but useful practical case in the following.

Magnetic Circuit Analysis

The analysis of magnetic devices generally begins with rough one-dimensional approximation techniques often called *magnetic circuit analysis*. Magnetic circuit analysis is reasonably cheap and speedy, and forms the main analytic tool of the practiced designer. It has survived the first quarter century of the computer era quite comfortably; no doubt it will continue to do so. In fact, this chapter is largely devoted to the proposition that *sophisticated analysis should always be used as a refinement and check on simpler calculations*, which in the present context really means that magnetic circuit analysis ought to be employed first to find out which cases should be subjected to thoroughgoing computational scrutiny by finite element methods.

To perform magnetic circuit analysis, the device to be analyzed is subdivided into segments of a more or less regular form, so chosen that the flux density in it may be expected to be more or less uniform. The magnetic field in each segment is assigned a fixed direction on the basis of experience and past practice. Each part is then viewed as a component with known reluctance, and a simple relationship (magnetomotive force equals the product of flux and reluctance) sometimes called the "magnetic Ohm's law" is used to describe each piece. The entire device is regarded as a collection of such finite segments, whose interconnected flux paths may be thought to form something analogous to an electric circuit. Mathematically, the resulting equations closely resemble those of dc electric circuit analysis.

Magnetic circuit equations are very simple and usually easy to solve with pencil and paper. Only in rather complex cases is recourse to a computer, or even a calculator, necessary. On the other hand, writing magnetic circuit equations for any device requires a close acquaintance with the physical behavior of the device, so that magnetic circuit analysis is really effective only when guided by intimate familiarity with the expected solution. Even in the hands of experienced designers, quantitative results obtained from magnetic circuit analysis are not very reliable; but they are generally fast to obtain, easy to interpret, and reasonably good guides to *comparative* performance of more or less closely related design variants.

A Gapped-Core Reactor Problem

To illustrate the principles involved, a comparatively simple problem will be examined. Fig. 1 shows a rough sketch of a *gapped-core reactor*, a simple magnetic device which consists essentially of an iron core (in principle somewhat similar to a transformer core) surrounded by a winding. The object of this device is to provide extra inductance in power circuits, by storing energy in magnetic form for part of an ac cycle and delivering it to an external circuit subsequently. In order to keep the total stored energy high, and also to ensure that it appears as a more or less linear circuit element, the reactor is endowed with a sequence of air gaps in its core. Thus the principal portion of the flux excited by the winding circulates around an iron core subdivided by air gaps. The design sketched in Fig. 1 is

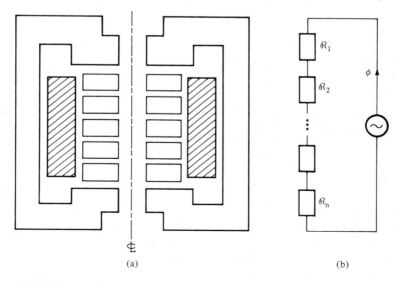

(a)   (b)

Figure 1. (a) A gapped-core reactor, an essentially axisymmetric iron structure surrounded by a winding. (b) Its magnetic circuit representation.

essentially axisymmetric. Its central portion, an iron limb surrounded by the winding, consists of annular iron rings separated by nonmagnetic spacers and held together by a large nonmagnetic central bolt. The bolt is not shown in the sketch, for the sake of clarity. The air gaps, which are actually not filled with air but with nonmagnetic spacer material, are quite numerous, perhaps ten or twenty altogether. The winding may consist of a single coil or several separated coils, depending on the voltage rating (and hence the insulation requirements) of the device.

A simple magnetic circuit model of this device, sufficient for a first analytic assessment, can be constructed quickly and easily if the following assumptions are made:

1. The energy stored by the reactor resides entirely in the air gaps; the iron core serves essentially as a flux-conductor that connects the air gaps. The energy stored in the iron may be neglected, and the iron may be considered infinitely permeable.

2. The flux crosses each air gap in a purely vertical direction, and the flux density is uniform throughout each air gap. In other words, flux lines are straight, parallel, and vertical.

3. The flux surrounding the winding itself, which does not pass through the iron core, is negligibly small. (In the conventional trade argot, "there is no leakage flux".)

The first assumption guarantees that the circuit "wiring" is made up of perfect, reluctance-free, flux paths between circuit elements; the second, that the circuit elements can be assigned values easily; and the third, that the "insulation" between circuit elements is perfect. Each circuit element (which should be called a *reluctor*, but never is) has a specified reluctance, and the circuit is driven by a source of magnetomotive force equal to the ampere-turns product of the winding.

Provided that the three key assumptions are not too far wrong, the magnetic circuit analysis of this device is simple enough to be done on the back of a used envelope, with the stored energy evaluated to sufficient accuracy to narrow the range of possible designs to a very few. Beyond that point, however, more refined analysis is required to compute the leakage fluxes and their associated forces, to estimate the iron losses, or to inquire after the flux distribution in the iron. Such more refined analyses are on occasion performed by a second level of magnetic circuit analysis, in which a more complex circuit is built. Leakage paths are then assumed and assigned reluctance values, and the iron core is broken up into portions which are assigned uniform, but not necessarily equal, flux densities. However, refinement in this direction is not common, because the key ingredient of magnetic circuit analysis still remains: the flux in every portion of the circuit is assumed to have a direction known in advance, only its magnitude remains to be determined. This fundamental assumption is generally not valid, except perhaps in very well known devices for which detailed test data exist. The usual next level of investigation therefore involves removal of this assumption, through finite element analysis.

Two-Dimensional Finite Elements

Viewed from the perspective of the experienced magnetic circuit analyst, finite elements could be regarded as generalizations of magnetic circuits. Where each magnetic circuit element is strictly one-dimensional (the flux direction in it is defined from the outset, and only the magnitude remains to be determined), the triangular finite element is two-dimensional, with both the magnitude and the direction of flux density unknown until the solution is actually produced.

To illustrate the additional information gained by moving to the next level of sophistication in analysis, consider the gapped-core reactor in Fig. 2, which shows the winding and air gaps of an actual reactor design. Only one-quarter of the device appears in Fig. 2, for reasons of symmetry. This particular reactor is designed for EHV (extra-high voltage) use, so that the winding consists of distinct layers separated by insulation. The layers are connected in series, with the low-voltage end of the winding nearest the

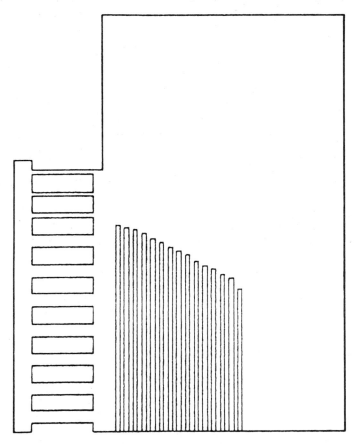

Figure 2. Winding and air space of EHV gapped-core reactor. The winding shape is chosen for insulation reasons.

center of the structure. The curved winding shape is dictated by insulation considerations, to provide additional clearance at the high-voltage (outer) side. In fact, the electromagnetic design of this device really consists of two essentially independent but interlocking designs, magnetic and electric.

Several different levels of refinement may be chosen for the field analysis of this reactor. Initially, the iron may still be considered infinitely permeable, but the assumptions of negligible fringing and leakage may be removed. After the design has been refined to some extent, the true nonlinear character of the iron may be gradually introduced in subsequent analyses.

Finite Elements in the Reactor Problem

Finite element analysis of the gapped-core reactor can be carried out at several levels of refinement. For the first level of finite element analysis, the winding should probably be considered as homogeneous, without regard to its layered structure, but the air gaps should be modelled in some detail. Such a model will provide accurate results for the stored energy (and hence the inductance) of the reactor but will not give detailed information on the field structure near the individual winding layers. It is therefore an excellent tool for overall design planning, but not much good for calculating winding forces, and of no use at all for determining the voltage distribution between layers. However, such a simple finite element model will be easy to construct and cheap to investigate. Two or three hundred triangular finite elements will usually suffice to model a structure such as that in Fig. 2; with an interactive analysis system like MagNet, the total working time will be of the order of perhaps half an hour or an hour. A great many more elements will not be worthwhile simply because accuracy will be limited by the remaining physical assumptions, not by the finite element technique.

At a second level of finite element sophistication, successive refinement may be carried out in one of several directions. For example, the designer may wish to introduce detailed modelling of the winding structure, by actually representing the winding as a set of individual layers. Doing so will permit not only the total inductance of the reactor, but also the self and mutual inductances of the individual layers to be determined. In the reactor shown, there are 15 separate layers, all connected in series. Taking the layers individually, an inductance matrix of order 15 results when all the mutual and self inductances have been found. The voltage distribution between layers can be determined readily once the inductance matrix is known. Coordination of electric and magnetic designs can then proceed. In an EHV reactor, this coordination of designs is vitally important. The main explicit specification of this device is magnetic: the reactor is to have a certain specified inductance and a certain minimum value of Q. However, the design possibilities are very strongly circumscribed by the intrinsic specification that voltage gradients in the internal insulation space must remain within reasonable bounds. Thus the high-voltage insulation requirements are in fact a major determinant of the size, and hence the cost, of the device.

A triangular finite element mesh suitable for analysis at the second level is shown in Fig. 3. The winding layers are shown in far greater detail than would be required for determination of air-gap flux densities and total stored energy. This model comprises 1044 triangular elements, with 619 nodes. For electric field studies, an analogous but probably different model is required, with still finer detailing at the winding ends where high electric fields are expected to occur, but fewer elements in the air gaps where the electric field is very likely to be negligible. The total element and node numbers for the electric and magnetic models are similar, however.

The results obtained at each level of finite element analysis may of course be used to verify the assumptions entering into the next simpler

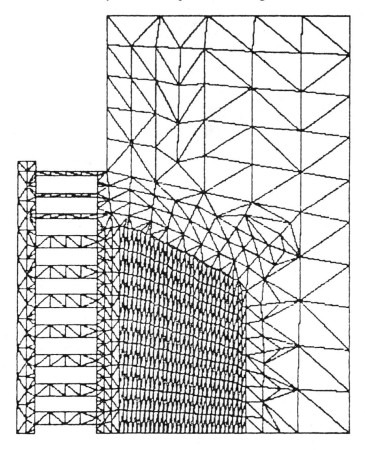

Figure 3. Finite element model of the gapped-core reactor, suitable for voltage distribution and insulation analysis. The winding layers are modelled in considerable detail.

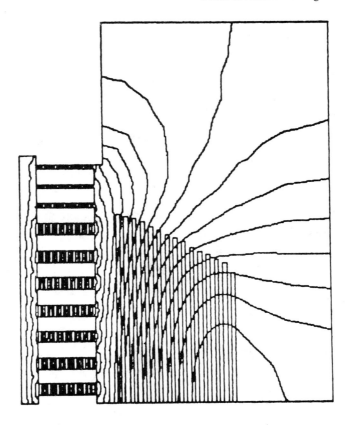

Figure 4. Magnetic equipotential plot for the finite element gapped-core reactor model of Fig. 3.

level, and indeed into the magnetic circuit analysis. For the model shown in Fig. 3, a magnetic equipotential plot is shown in Fig. 4. (Since this problem is treated as axisymmetric, the equipotentials are not flux lines; flux lines are obtained if the potential values are multiplied by the radius.) An enlarged view of the center portion of the reactor appears in Fig. 5. Several points of interest emerge from these plots.

There is evidently some not altogether negligible amount of fringing flux which makes its way down the center of the core, where the nonmagnetic main bolt is located. This flux may give rise to eddy currents, and an associated power loss, in the bolt. If the bolt is to be made of metal, Figs. 4 and 5 suggest that a still further finite element model may be worthwhile: one in which eddy currents are represented, so that their magnitude as well as the eddy current loss may be estimated. Such a model clearly does not need to be very detailed in the winding region, but it should pay due attention to correct representation of the bolt and the

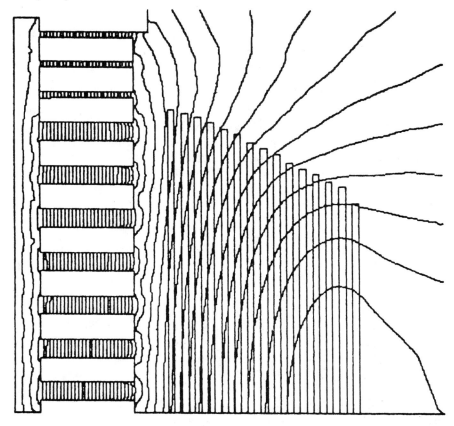

Figure 5. Magnetic vector equipotentials in the central portion of the gapped-core reactor model.

nearby air gaps. If the eddy current losses are substantial, the use of a nonmetallic bolt material may be indicated.

In the air gaps, flux lines are almost straight, for Fig. 5 shows only a small amount of fringing. It may be concluded that stored energy estimates obtained by magnetic circuit analysis are quite likely accurate in this case.

Within the winding region, equipotentials (and hence flux lines) are substantially straight, and nearly vertical, for the inner layers. Most of the turns in each layer therefore link about the same amount of flux. At the high-voltage end of the winding, however, a larger number of flux lines cross through the layers. As a result, there may be a significant unevenness in voltage distribution along a layer. If insulation problems are likely to be very important, there may be merit in representing the winding not only as a set of distinct layers but also dividing each layer into several series sections, so as to obtain an improved circuit model of the winding for studying the voltage distribution in the winding.

There are evidently quite a few different ways of modelling the gapped-core reactor at the second level of complexity, each model being useful for a different purpose. Not all models need necessarily be constructed for every, or indeed any, proposed design. However, prior to actually building an expensive prototype, a still more complete performance prediction of the reactor may be sought by means of a third level of finite element analysis. This level involves dropping the assumption that iron is infinitely permeable, and substituting a reasonably accurate saturable magnetic material model. For reasonably detailed modelling of the iron shapes, a few hundred additional finite elements will be needed. In other words, the total number of triangular elements at this level is about 1000 if the winding is not represented in fine detail and may rise to 1500 or so if it is.

Problem Complexity and Flexibility

The gapped-core reactor is a device of moderate complexity, yet, as seen above, a truly detailed model useful for every purpose can easily require many thousands of finite elements. Thus even two-dimensional problems can easily stress the capabilities of large-scale analysis systems. The obvious way of countering the difficulty is not to build a model of the reactor with everything modelled in detail but to construct instead several models whose overall complexity is moderate, but which are highly refined in a single aspect each. This kind of selective refinement has been discussed in some detail earlier.

Several models of moderate complexity are usually less costly in computing resources than a single large model. The very best sparsity-exploiting simultaneous equation solvers available are those using gradient methods, which are found experimentally to involve computing times proportional to the three-halves power of the number of equations. Other conventional solution techniques, such as band-solvers, require time proportional to the square of the number of equations. In large problems the number of equations to be solved is roughly one-half the number of triangles. (It can be shown that the number of nodes in a first-order finite element mesh cannot be less than half the number of triangles, and it approaches this limiting ratio fairly rapidly as the number of elements increases beyond a few hundred.) Thus, doubling the number of elements in a model results in a threefold or fourfold increase in computing time, usually accompanied by a corresponding increase in man-hours. The selective refinement approach suggested above is thus the correct one: the several aspects of a device should be investigated using several models of moderate complexity, not by building a single model complex enough to do everything.

The objection to constructing a sequence of models lies in just that: this approach requires a sequence of models, each of which must be specified. In fact, creating a multiplicity of models need not involve a great deal of additional work, provided some advance planning is done. In the gapped-core reactor problem, for example, it is wise to build an initial model in which the winding is represented fairly crudely. Another model, similar but with a better winding model, can be created by editing procedures which involve removal of the winding and its replacement by another but do not require rebuilding of the model. Since it is necessary in any case to create a detailed model of the winding, the main additional effort involved in multiple models is the creation of an additional crude mesh. However, the work involved in doing so is generally quite small.

Subproblem Analysis

Subproblem analysis means to the magnetics designer what the term *complexity aggregation* signifies to the control engineer: the replacement of an aggregate of complicated details by a simpler shape, whose behavior is similar as far as the external world is concerned. The entire problem is thought to be composed of numerous subproblems, each of which is analyzed in detail. The fundamental assumption involved is that the various subproblems only affect each other mildly, so that piecemeal analysis makes sense. The classical Carter coefficient may have been the first such aggregation to become widely known. Others, however, are used daily. One such subproblem analysis will be sketched out here, by way of example.

The Butt-Gap Problem

Transformer cores, as well as the iron portions of many other magnetic devices, are conventionally built up of stacked laminations in order to minimize eddy current losses. The laminations are rarely of a shape which will permit flux paths to close within one and the same metal part; laminations can easily be cut to such shapes, but difficulty is encountered in attempting to place windings on closed iron paths. It is therefore normal to make each lamination span part of the magnetic circuit, the remaining flux path being taken care of by another. The pairs (occasionally larger sets) of laminations butt up against each other edge to edge, and unavoidably leave a small air gap. The effective gap is reduced by making two (on occasion more) kinds of laminations, with the flux path broken in different places, then stacking them so as to have butt joints occur only in alternate laminations. While this procedure reduces the effective gap, it does not eliminate it. Just how large a residual effect the many little gaps do have is a question of considerable interest to designers.

The stacking of joints in alternate laminations then means that a laminated iron core must have an appearance near the gap joint somewhat as shown in Fig. 6. The butt gaps are assumed to be sharp and rectangular, although they are not necessarily so in reality. Similarly, the laminations

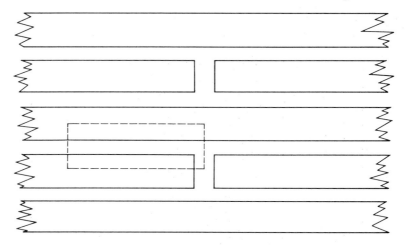

Figure 6. Part of a packet of laminations, showing butt gaps in alternate layers. There are three symmetry planes near each butt joint, so analysis can be confined to the region shown in dotted outline.

themselves are assumed to be smooth and flat, and separated by flat sheets of nonmagnetic material, assumptions which again might be questioned. Nevertheless, under these assumptions, and ignoring events at great distances, the center-plane of each lamination must be a plane of symmetry, and so must the joint center-plane. Consequently, only the dotted region of Fig. 6 need actually be analyzed, the remainder of the solution can be deduced from symmetry.

The left-hand edge of the analysis region in Fig. 6 is neither a plane of symmetry nor a true boundary of the problem region. Theoretically, the region to be analyzed should extend infinitely far to the left. However, some distance from the joint the flux lines may be expected to be straight and horizontal, with uniform density within each lamination. It is therefore reasonable to approximate the situation by terminating the region at an arbitrary plane a few sheet thicknesses distant from the joint, as has been done in Fig. 6, and to assume that all flux lines cross that plane orthogonally.

The object of analysis in this case is to replace the set of butt joints, near which the flux distribution has some complicated shape, by a hypothetical butt joint which the flux lines are constrained to cross exactly at right angles. At the same time, the packet of laminations is to be replaced by a homogeneous material whose magnetic behavior is the same as that of the packet of laminations. Viewed in magnetic circuit terms, the complicated butt joint is to be modelled by a series reluctance; the problem is to determine its value. The approach adopted here is to solve the real problem, find the stored energy increase due to the gap, and then to set up the simplified problem (which is simple enough to require no computer) so as to duplicate the stored energy. The real and artificial situations will then be judged equivalent from an energy storage point of view. The resulting nominal gap value and the properties of the artificial magnetic material in which it is embedded may then be used to simplify transformer analysis by permitting the core to be modelled as a homogeneous material with a simple air gap, instead of the very complicated structure of laminations and gaps which it really is.

Incompleteness of Results

For a typical butt gap in transformer laminations, the flux lines behave as shown in Fig. 7. The problem as shown here formed part of a design study, in which a large set of similar problems was solved for various total flux levels, various butt gaps, and various interlamination spacings, for a

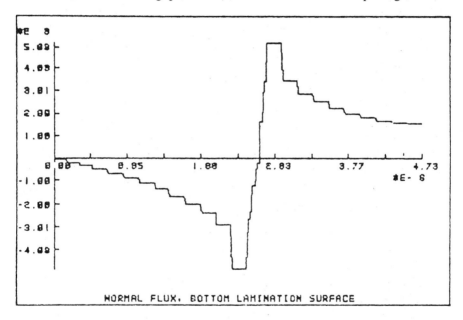

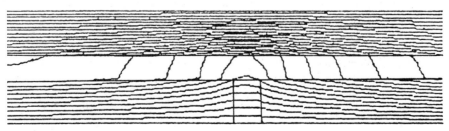

Figure 7. Typical flux-line behavior near a butt gap in a saturable material. The bottom plot shows the path taken by the flux lines, while the upper curve shows the flux density (in normalized units) orthogonal to the upper surface of the lower lamination.

particular type of saturable iron. The top and bottom lamination center-lines were required to be flux lines, and the total amount of flux between them was prescribed, thereby prescribing the average flux density in the equivalent homogeneous material. There is no particular difficulty about computing the stored energy in this situation, and the determination of equivalent smooth air gap width is not a problem.

As might be expected, the close proximity of the upper lamination means that flux lines do not all cross the butt gap directly. Part of the fringing flux at the joint side in fact crosses over to the neighboring lamination, raising the local flux density there. This phenomenon is especially marked when the iron is more or less heavily saturated, and it raises a practical difficulty which can at times become serious. Flux lines crossing the interlamination air space, as in Fig. 7, give rise to a vertical component of flux density. In other words, there exists a component of flux density normal to the lamination plane. Any eddy currents that may be evoked by this flux component will tend to flow more or less unhindered in the plane of the lamination, potentially giving rise to a significant local eddy current loss. The normal component of flux density at the lamina-tion surface is therefore of great interest. It is shown plotted (in normal-ized units) in Fig. 7.

Boundary Conditions

Electromagnetic fields in most cases really exist in an infinitely extending space. The Maxwell equations describe the electric and magnetic fields, and implicitly prescribe their behavior near material interfaces. However, there are rarely any true boundaries in an electromagnetic field problem, for only rarely are the fields truly bounded. In most cases, boundaries are imposed on problems in an approximate sense. An iron body in air, for example, does not bound the magnetic field. But since the permeability of iron is much higher than air, flux lines in air must be nearly orthogonal to the iron surface. If only the field behavior in air is of interest, a strict boundary is often established by assuming them to be exactly orthogonal. Such artificial boundary conditions occur in the butt-gap problem as for-mulated above, and reappear in many other cases.

In this section, some of the difficulties in formulating boundary condi-tions will be considered and possible solutions indicated, primarily by means of illustrative problems.

Accelerator Magnet Study

The simplest form of boundary condition to impose on any problem is one which really holds true, for there can then be no argument about its degree of approximation. In the main, such boundary conditions actually occur only along planes or lines of symmetry. Fortunately, the aesthetic sense of designers as well as the realities of fabrication technology imply that a great many devices possess one or more planes of symmetry.

One-quarter of the cross-section of a superconductive magnet, investigated as part of a design study preliminary to detailed design of a large high-energy particle accelerator, is shown in Fig. 8. The magnet is quite long (in the direction normal to the plane of the drawing) compared to its cross-sectional dimensions, and the circular hole in its center forms a long tunnel through which elementary particles are accelerated. The tunnel or bore is surrounded by a system of coils which occupy a nearly cylindrical space around the bore and maintain an essentially dipolar magnetic field in it. The bore and coils are encased in a large iron yoke.

The mathematical physics of the problem is simple enough, in principle at any rate. In the entire iron and air region, the magnetic field is adequately described by the magnetic vector potential **A**. Since the current density in the coils may be assumed to possess only a longitudinally

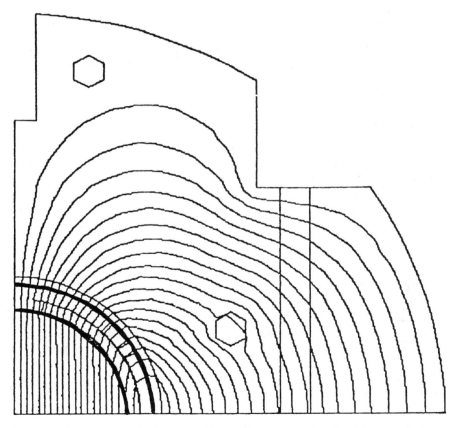

Figure 8. One-quarter of a large particle accelerator magnet. Particles travel along the center bore in a magnetic field maintained by superconductive coils, the whole being encased in an iron structure.

directed (i.e., into the plane of the paper) component, the magnetic vector potential has only a longitudinally directed component also. Everywhere in the iron as well as the bore and coils, **A** must satisfy the nonlinear Poisson equation

$$\text{curl } v \text{ curl } \mathbf{A} = -\mathbf{J}, \tag{1}$$

subject to suitable boundary conditions, as in the following. The vertical symmetry line bisects the coils, with exactly similar but oppositely directed distributions of the current density **J** on its left and right sides. It must therefore be a flux line, or more precisely, the separatrix curve which separates the flux lines closing about the right-hand coil sides from the flux lines closing about the left-hand coil sides. In translationally uniform problems, flux lines are lines of constant A, as discussed elsewhere in this book. A symmetry line is forced to be a flux line, and odd symmetry of the solution values is assured, by setting

$$A = 0 \quad \text{along the vertical symmetry line.} \tag{2}$$

The bottom edge of the drawing in Fig. 8 is also a symmetry line but differs from the left edge in one important respect: the current density in the coils is distributed similarly about the horizontal (bottom) edge and directed in a like sense above and below, but the currents are oppositely directed on the two sides of the vertical axis. Magnetically, the vertical axis is therefore a line of odd symmetry, while the horizontal axis is a line of even symmetry. Flux lines that encircle the upper half of the coil must also encircle the bottom half in a like fashion. Hence the flux lines must cross the horizontal plane orthogonally. In other words, the classical *Neumann* boundary condition holds:

$$\frac{\partial A}{\partial n} = 0 \quad \text{along the horizontal symmetry line.} \tag{3}$$

In finite element solutions, this type of boundary condition is *natural*, that is, it will be imposed by the solution process itself (in an approximate sense) without any need for the user to specify it explicitly.

The exterior edge of the iron yoke forms the remaining portion of the problem boundary. Here no useful symmetry conditions apply. On physical grounds, it is very likely that most by far of the magnetic flux lines will remain inside the iron. If it is assumed that none of the flux lines leaves the iron, one flux line must follow the iron boundary. This must indeed be the same flux line that runs down the vertical axis (the plane of odd symmetry) of the magnet structure, so the requisite boundary condition is of the *Dirichlet* type, and is homogeneous:

$$A = 0 \quad \text{along iron boundary.} \tag{4}$$

It is important to note that this boundary condition, unlike the other two, is approximate, based on the designer's prior knowledge and expectations arising from experience. Expectations can be wrong, and it may at times be wise to check them. A check can be obtained in this case by enlarging the problem region to include some outside air and requiring the outside edge of the air region to be the flux line $A = 0$. If a flux line does indeed follow the iron surface as expected, this fact will be immediately evident in the solution. How large an air region to add and where to place boundary conditions may be better evident from the next example.

A Recording Head Problem

Many problems in magnetics involve very distant boundaries. For example, the magnetic field of single-turn coil placed in air is known to fall off quite rapidly at large distances, but still it has no well defined limits. Even when a problem is well bounded, it often happens that the boundaries enclose quite large vacuous spaces, in which the magnetic field is weak and not of great interest to the designer. These spaces need not usually be modelled in any great detail, because they rarely have a substantial effect on the solution. At least part of the space, however, needs to be modelled all the same. Placement of artificial boundaries and mesh grading in the boundary region are therefore of some importance to problem formulation.

One-half of a horizontal magnetic recording head appears in Fig. 9. The recording medium is very thin, hence not visible in the full drawing.

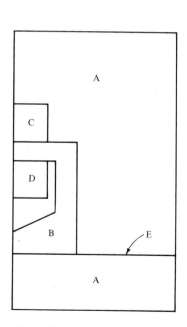
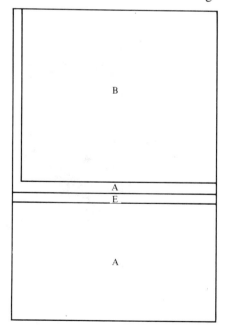

Figure 9. Left: Outline sketch of one-half of a magnetic horizontal recording head, showing the head shape and coil placement. Right: Greatly enlarged view of the pole B and magnetic oxide film E.

Similarly, the interpolar gap of the head (at the left edge of the drawing) is very small and invisible. Both can be discerned, however, in the enlarged plot on the right. The recording head structure is symmetric about the left edge of the drawing, and the coil currents have even symmetry about this plane. As discussed in detail in the previous example, the left edge must therefore be a boundary of (homogeneous) Neumann type. This boundary condition is exact, not approximate, because it applies at a symmetry plane which forms a boundary in a natural fashion. However, there are no other symmetry lines in this problem so the remaining boundary conditions will have to be imposed by artificial approximation.

Approximate boundary placement and boundary conditions are best based on a qualitative knowledge of the solution to be expected. To aid the imagination, Fig. 10 shows a solution for the magnetic field of the recording head of Fig. 9. Although there is not even a hint of symmetry

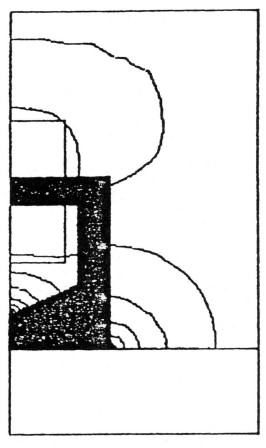

Figure 10. Flux lines near the recording head of Fig. 9. In order to show the leakage flux clearly, so many lines have been drawn as to color the iron parts essentially solid.

about any horizontal plane, the behavior of flux lines is topologically similar to that encountered in the accelerator magnet problem. There must exist a separatrix curve which partitions the drawing of Fig. 10 in such a way that all flux lines above the separatrix encircle the upper coil side and all flux lines below it encircle the lower. This separatrix must begin somewhere on the left edge of the drawing and terminate somewhere on the right edge, and the limiting curve in each of the two families of flux lines must follow the edge of the discretized problem region. If the value of vector potential **A** is taken to be zero along the separatrix, as a matter of convenience, then the flux lines that follow the edges of the drawing must correspond to zero vector potential also. In consequence, a well defined problem is obtained by insisting that the upper, lower, and right edges of the drawing all have $A = 0$.

How far from the central region of interest such an artificial boundary should be placed is often a matter of guesswork. The guesses can be improved by noting two significant facts. First, many magnetics problems do not require that the correct solution be produced over the entire region modelled, but only over some central portion of particular significance. The recording head example is a particularly good example, in that the quantity of prime interest is the magnetic field in the recording film near the head gap, which is unlikely to be much affected by the accuracy with which the leakage field at great distances is modelled. Secondly, every magnetics problem must contain a closed set of currents, so that its field at a sufficient distance is essentially that of a dipole. The field of a dipole falls off rapidly with distance, so that it is very rarely necessary to carry modelling beyond a distance one or two times the major dimension (in this case, head size) of interest.

Perpendicular Recording Head

Another recording head problem is illustrated in Fig. 11. In this case, the head is one designed for perpendicular recording; it possesses no symmetry planes at all. To choose an appropriate set of boundary conditions is thus again a matter of assessing what the questions of importance may be and what the solution may look like in reality. The latter is largely a matter of experience, but experience can again be aided by computation. If no other guideline exists, it is always possible to encase the problem in a very large air space indeed, to compute a rough solution, and to deduce from the rough solution what the qualitative behavior of a correct solution is likely to be. To assist in discussion, a reasonably accurate solution of the perpendicular head problem appears in Fig. 12.

In common with the problems examined previously, Fig. 11 contains two coil sides, with equal but oppositely directed currents. As before, the flux lines are therefore separable into two groups: one whose members all encircle positively directed currents and one whose members all encircle negatively directed currents. Because the iron shapes in Fig. 11 are asym-

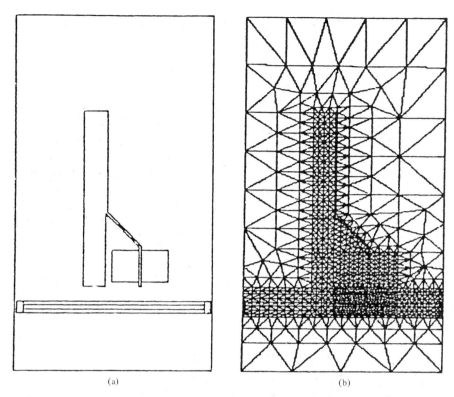

Figure 11. A perpendicular recording head. (a) Overall sketch, showing the key parts. Recording is caused by currents in the coil placed on the narrow (high flux density) iron leg. (b) Mesh of 1750 finite elements.

metric and complicated, it is not at all obvious where the separatrix curve lies that forms the boundary between the two. In fact, it can be found only after some study of the solution, Fig. 12(a), which shows where the flux lines are located. The separatrix curve is shown separately in Fig. 12(b), for the sake of clarity.

In every problem containing zero total current (i.e., wherever the currents in all coil sides or conductors balance), a separatrix curve must exist, which partitions the problem space in such a way that all flux lines in the same partition encircle the same conductor and no flux line ever crosses the separatrix. If there are numerous coil sides or conductors, the separatrix may in fact be a multi-branched curve. Its exact placement is not the issue at hand here; rather, the matter of importance is that such a curve exists and must always exist whenever the total currents balance. Because the flux lines in such a problem must close entirely within the problem region, some or all branches of the separatrix must touch the exterior boundary. Flux lines must exist which lie parallel to the separatrix and very close to it. But since flux lines must be closed, the exterior

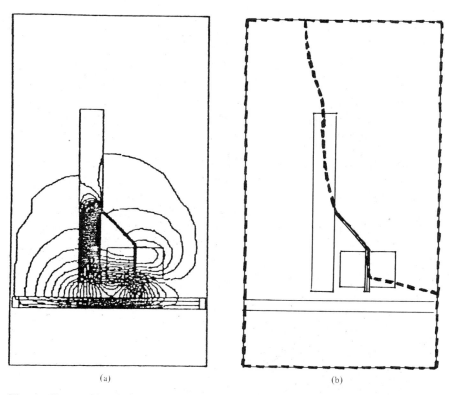

Figure 12. (a) Flux plot for the perpendicular recording head. (b) The dotted separatrix curve partitions the problem space between flux lines with clockwise and anticlockwise circulation.

boundary must itself be part of the separatrix wherever the separatrix touches it. Hence it is possible to establish a boundary condition rule: *every closed system of currents must be surrounded by a flux-line boundary and crossed by a separatrix curve*, a flux line which joins the exterior boundary.

In the accelerator magnet problem, balanced coil currents exist, even though they do not appear in the quarter-plane model. Since the vertical axis is a line of odd symmetry in Fig. 8, it was argued to be a flux-line boundary on symmetry grounds. However, an equally good case could be made on the more general basis derived from the perpendicular recording head example. Were the full accelerator magnet to be analyzed, not merely one-quarter of it, the problem would contain balanced currents. A separatrix curve must then exist, and the entire problem must be bounded by a flux line. The separatrix curve, however, must be located on the axis of symmetry, as may be seen from the following argument. Were all currents to be reversed, all flux lines would reverse but their distribution would remain otherwise the same. In other words, reversal of current should change the field solution into its own negative; and in this case the separatrix curve must remain the separatrix still. But that is only possible if the separatrix coincides with the symmetry line, so that it is its own mirror image.

Electric Machine Problems

Electric machines generally involve complicated but highly repetitive geometric structure. The repetitiousness frequently makes it possible to model and solve only a part of a machine problem, modelling only one pole pitch or even only half a pole pitch of a multipolar machine.

When a machine is operated under no load, only field currents but no armature currents flow. So long as the field winding is symmetrically arranged, the flux distribution will be symmetric about both the pole axis and the interpolar axis. Since in this case there exist two symmetry planes half a pole pitch apart, only half a pole pitch need be modelled. This fact may be verified by example, by considering the full pole pitch model of a turbogenerator shown in Fig. 13.

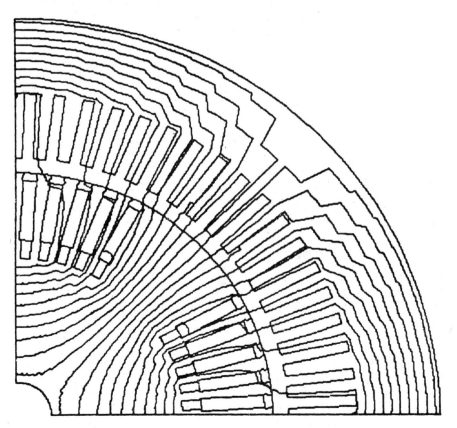

Figure 13. One pole pitch of a four-pole turbogenerator running unloaded.

Since the machine is not loaded, no current flows in the turbogenerator stator slots, but the field winding placed in the rotor slots carries current. Both radial edges in Fig. 13 are lines of even symmetry, for each bisects one side of a field coil. Both radial lines must therefore be homogeneous Neumann boundaries, lines of mirror symmetry, which flux lines cross orthogonally. Because such boundaries are natural in the finite element method, there is no need to specify any constraint along these edges when building finite element models; it is sufficient simply to leave the potentials to go whither they will. In this machine, flux lines are partitioned by separatrix curves which coincide with the polar axes; they are closed by the outside edge and shaft of the machine. When building the model, it is therefore necessary to enforce a flux-line boundary along the shaft edge as well as along the outside boundary, by setting $A = 0$ there.

The model of Fig. 13 evidently encompasses twice as large a problem region as is actually necessary. The solution shown makes this fact amply clear: the pole axis is obviously a line of odd symmetry, the partitioning line between flux lines which link the left and right coil halves, respectively. Were the stator winding excited and the field current reduced to zero, a similar result would obtain, except that the line of odd symmetry would line along the interpolar axis, and the line of even symmetry along the pole axis. With both rotor and stator energized, as would be the case for operation under load, neither axis would remain a line of symmetry, because the two fields would add and subtract in alternate half pole pitches. Analysis of a full pole pitch is therefore indicated, as shown in the following.

Periodicity Conditions

The lack of symmetry that results when both armature and field of an electric machine are excited need not cause concern for the analyst. To illustrate this situation, Fig. 14 shows the full magnetic field solution of a four-pole fully compensated dc machine. It is immediately obvious that there is no radial line anywhere which represents an axis of either even or odd symmetry. The shift of rotor flux off center, with its corresponding shift in pole and commutating pole fluxes, is clearly visible.

Examination of the flux distribution of Fig. 14 does reveal, however, that it is clearly periodic: there are four identical regions partitioned by a four-branched separatrix curve, whose branches cross at the center of the machine shaft. The fluxes circulate alternately in clockwise and anticlockwise directions; correspondingly, the vector potential values are alternately negative and positive in successive quadrants. Values of A therefore repeat at intervals of two pole pitches and reverse in sign at intervals of one pole pitch:

$$A(r,\theta) = A(r,\theta+\pi) \tag{4}$$

and

$$A(r,\theta) = -A(r,\theta+\pi/2). \tag{5}$$

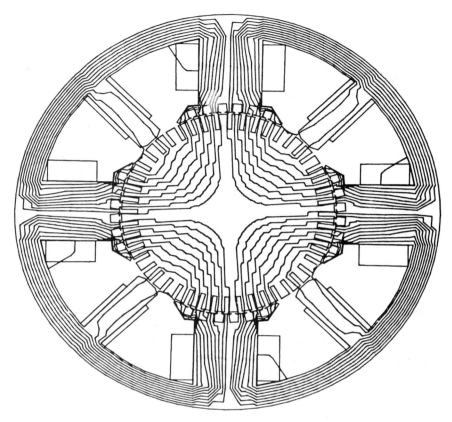

Figure 14. Flux distribution in a compensated direct-current machine under full load.

If two radial lines are drawn, exactly one pole pitch apart but otherwise arbitrarily located, equation (5) states that the vector potential at a given radial position on one line will be exactly the negative of the vector potential at the corresponding radial position on the other. Neither value, however, is known. Such boundary conditions are known as *periodicity conditions* and occur quite commonly not only in electric machines but various other repetitive or periodic structures. Because the fields are repetitive, one cell or unit of the repetitive pattern is all that needs to be analyzed. It is in fact not even necessary for the cell to be bounded by radial lines in the case of an electric machine. In the dc machine, for example, any curve will do as a cell boundary, provided it reaches from the machine axis to the outside perimeter, and provided the cell is bounded by another curve of like shape but translated one-half pole pitch.

Periodicity conditions represent a special case of the more general *binary constraint* which binds together values of potential at points P and

Q by the relationship

$$A_P + k_{PQ} A_Q = R_{PQ}. \tag{6}$$

Such linking together of values is easily accomplished in finite element models, by linking together appropriate nodal potentials through equations of the above form. Provision is made for fairly general binary constraints in the MagNet analysis system, and binary constraints in one form or another appear in almost every available CAD system. At the very least, periodicity conditions should be provided in any system that deserves to be taken seriously.

Geometric Shape Alteration

Magnetic device design may from time to time require investigation of comparative behavior under geometric shape changes, indeed many design problems have shape optimization as their primary goal. In such cases it is necessary to solve and compare sequences of problems defined on different mesh models. The simplest way of accomplishing this objective is to alter not the mesh, but the materials used in the mesh, as illustrated in the following.

Butted Transformer Laminations

In the above, mention has been made of the problem of butt gaps in packets of transformer laminations, with a sample flux distribution shown in Fig. 7. That flux distribution shows the principal features of interest—the general way in which flux lines cross or circumvent the gap in the butt joint, the manner in which normal components of flux density arise, and the depth to which normal fluxes penetrate into the lamination. It exhibits the relevant physical phenomena in a thoroughly satisfactory way but does not fully satisfy the designer's need for quantitative information. Knowing the flux densities and estimated losses in a particular situation is all well and good, but how does the normal flux density vary with gap size? with saturation level? as a function of interlamination insulation? To answer these questions, extensive computation is required on essentially similar problems.

In the study on butt joints from which Fig. 7 is taken, four different butt gap sizes were investigated, for each of several different interlamination gaps and for several saturation levels. To avoid the bother of building a large number of separate meshes, a small number of models was constructed, with the gap region in each model made up of layers containing nominally different materials. The structure of such a model is clearly visible in Fig. 15, which shows not only the flux distribution in the laminations but the interfaces between different materials.

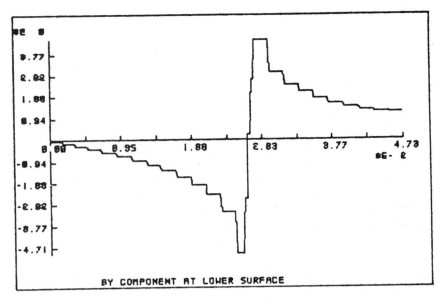

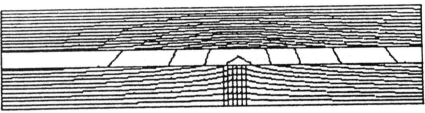

Figure 15. The butt gap between transformer laminations modelled as a sequence of distinct materials.

With the gap region modelled as a five-layer sandwich of distinct materials, each of thickness equal to 5% of the lamination thickness, some dozen or more distinct cases can be investigated without change of geometric model. It suffices first to identify four of the materials in the sandwich as iron, one as air; then three as iron, two as air; continuing, to create butt gaps of 5%, 10%, ..., 25% widths (relative to lamination thickness). Each can be combined with excitations of several different levels to give the required distinct combinations of saturation level and gap width. Examination of Fig. 15 shows clearly that in this particular case, the leftmost pair of material slices has been identified as iron, and the right-hand set of three as air, giving a 15% gap.

This method of approach would be valuable if it merely saved working time. However, it has a second, and perhaps more significant aspect: virtual elimination of discretization error from comparative evaluations. When two solutions have been computed on the same mesh model, the discretization error in their difference can be expected to be much smaller than the error in the difference between two solutions computed on different meshes. When several results are to be compared, and differences between them calculated, the use of a common mesh model is to be recommended.

Precisely this procedure was used in the vertical recording head study briefly described above, to allow various sizes of gap to be introduced between the head and the recording medium. In that case, the amount of magnetic material was actually reduced a small amount, by being converted to air in the analysis. But since the gap in real recording heads is quite small compared with head dimensions, this approximation is a good one in preliminary work.

Boltholes in a Magnet Structure

A major part of the design study for the accelerator magnet of Fig. 8 focussed on the effect of boltholes, required to secure the magnet structure mechanically, on field uniformity in the beam region actually occupied by particles. This investigation followed the same technique as above: a single mesh model was built, and materials re-identified to effect the necessary changes.

When an iron bolt is placed in a packet of iron laminations, a substantial air space generally remains around the bolt; it is quite difficult to achieve a close fit in laminated materials. The size of this air gap is subject to variations in fabrication processes (especially lamination stacking and punching), and it is not easy to predict. From the designer's point of view, all that can be guaranteed is that the effect of bolt clearance will be greater than that obtained with zero clearance (i.e., no air at all) and less than that resulting from an empty bolthole. The design problem therefore reduces to choosing bolthole locations so that the field in the magnet bore is affected little enough by the switch from one extreme (empty bolthole) to the other (no bolthole). This switch is accommodated in a single mesh model by re-identification of the material in the bolthole as either air or iron. The results of such a comparison are shown in Fig. 16, which contains two flux plots overlaid on each other, one with and one without the bolthole present. Evidently, the effect of the upper bolthole is very minor, as one might well expect from a hole very near the dipole axis and hence in a region of low flux density. On the other hand, the right-hand bolthole changes the flux distribution in the iron yoke quite substantially but has a surprisingly minor effect within the magnet bore.

Investigation of the accelerator magnet structure included also consideration of varying bolthole sizes and changes in placement. Some of the different combinations possible are hinted at by Fig. 17, which reproduces the flux plot of Fig. 8 but shows also the alternative bolthole placements provided for by the mesh model in use. Here, as in Fig. 16, the flux plot is accompanied by a curve giving the vertical component of flux den-

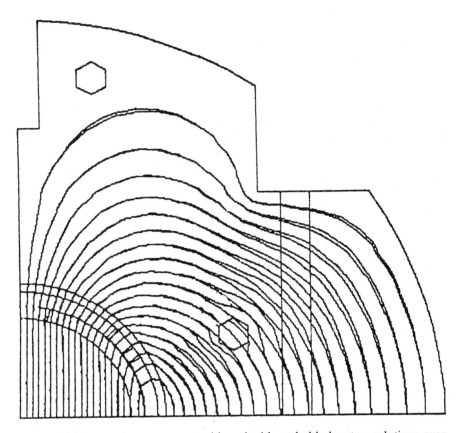

Figure 16. The accelerator magnet with and without boltholes: two solutions overlaid on each other.

sity along the drawing horizontal axis, from the bore centerline to the inner surface of the coil. This field component is visibly uniform to within the resolution of the drawing. The horizontal flux density component along this axis is zero for reasons of symmetry and is not drawn.

The creation of a mesh model in this case requires careful planning and involves more complication than would be the case for simple geometries like the butt-gap problem. The mesh on which Figs. 8, 16, and 17 are based comprises 1115 elements and is shown in Fig. 18. The boltholes, which are of course round in reality, have been modelled by hexagons, and provision is made in the mesh for material re-identification by including suitable hexagonal patches. This relatively crude representation of circles is permissible, despite the fact that the problem inherently seeks high accuracy, for two reasons. First, identical meshes are used, so that discretization error, even if not negligible, will have the same character in each case and will therefore tend to cancel. Secondly, the object of the study is to find such bolthole placements as to make the field in the bore

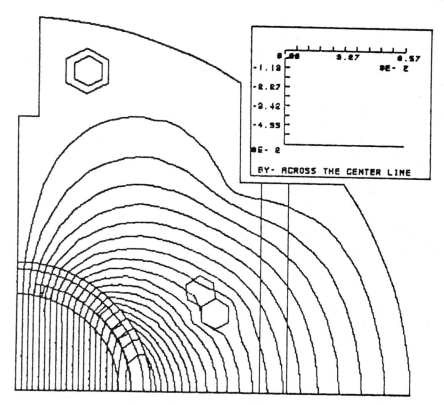

Figure 17. The accelerator magnet, showing alternative bolthole placement and size. The solution refers to the lower hole position and the smaller upper hole size.

substantially independent of the presence of the hole. It may be assumed quite safely that insensitivity to hexagonal holes implies insensitivity to round holes also!

In addition to providing the possibility of geometric shape alteration, the material re-identification technique may of course be used also for trying out a range of material magnetic characteristics as part of the design process. Furthermore, it is very useful in estimating forces or torques by the virtual work method. In this method, two solutions are produced, alike except for a small movement of the object on which the force or torque is to be calculated. The total energies for the two situations are determined, and the force or torque is then found as the ratio of the change in stored energy to its associated mechanical displacement.

Reducing Three Dimensions to Two

Real life is three-dimensional, and real CAD today is two-dimensional. There certainly exist practical problems which can only be solved in three

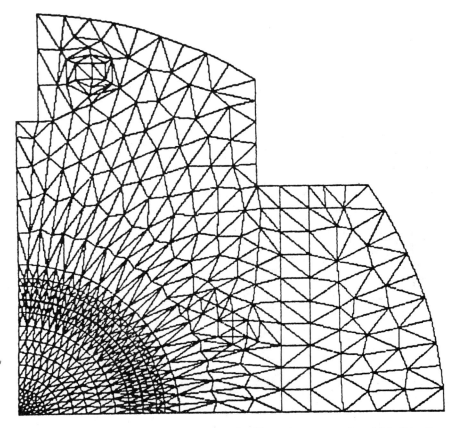

Figure 18. Mesh used for the magnet study. Note the manner in which the alternative bolthole locations and sizes are incorporated in the mesh.

dimensions, but a great many truly three-dimensional design problems will yield to two-dimensional approximations. Not only is three-dimensional analysis expensive; it so quickly overwhelms the user with data that the answers become unusable by virtue of their sheer volume. For the immediate future, a main use of three-dimensional field solutions will be subproblem analysis, leading to usable two-dimensional or quasi-two-dimensional models for analysis of the full device.

Ventilating Ducts in Machines

Electric machines of all sizes are built with both rotor and stator slotted, so as to make room for the windings. Large machines, such as turbine generators, are conventionally built with the stator core not stacked uniformly along its whole length, but rather with the stator built of lamination packets a few centimeters thick and separated by ventilation ducts a few millimeters or a centimeter broad. Consequently, the length of the iron packet forming the stator is not simply the axial length of the machine, but some shorter length. The problem now facing the analyst is that a particular amount of flux which crosses the stator surface must give rise to a higher flux density than would have been expected in the stator were it made of homogeneous material.

A first, and simple, approach to this problem is through modification of the stator magnetization curve, on the basis suggested by the following argument. For a given magnetomotive force, the actual stator carries less flux than would be the case for a stator of the same length but without ventilating ducts. Provided the flux density lies in the transverse (r-θ) plane everywhere in the stator iron, the iron and the ventilating ducts may be replaced by a homogeneous, composite fictitious material F, whose permeability is lower than that of the real stator iron S but substantially exceeds that of air. This replacement is similar to that of a laminated iron packet (e.g., a transformer core) by an artificial, geometrically anisotropic material, discussed earlier in connection with the modelling of magnetization curves. In fact, the present problem may be viewed as a simple special case of the former, since longitudinal flux has been assumed absent in the machine. If the fictitious material F is assumed to fill the same space as is actually occupied by stator iron and ventilating ducts, then the magnetic field $\mathbf{H}$ in the fictitious material is identically the same as it is in the stator iron packets, because $\mathbf{H}$ is created by coil currents which do not depend on the surrounding material. On the other hand, the flux density $\mathbf{B}_F$ in the fictitious material must be such as to produce the correct total flux in the machine. It will therefore be related to the magnetic field $\mathbf{H}_i$ in the iron by

$$\mathbf{B}_F = s(\mu_{iron} - \mu_0)\mathbf{H}_i + \mu_0\mathbf{H}_i, \tag{7}$$

in accordance with the discussion presented earlier. Here the dimensionless number s represents the proportion of stator space actually filled with iron,

$$s = \frac{d_{iron}}{d_{iron} + d_{air}}, \tag{8}$$

where d_{iron} is the axial total length of all the iron that comprises the stator packets, and d_{air} denotes the axial length of all the air, including the ventilation ducts as well as any interlamination insulation spaces in the laminated packets themselves. In other words, s represents a *fill factor* for the stator, a generalized stacking factor which includes both ventilation duct spacing and the lamination stacking. The analysis of a machine including ventilation ducts should therefore employ a mean magnetization curve which has the magnetic field H_F identical to that of the curve given for the iron alone, $H_F = H_{iron}$, but a different B_F. Equation (7) may be rewritten in the more convenient form

$$\mathbf{B}_F = s\mu_{iron}\, \mathbf{H}_{iron} + (1 - s)\, \mu_0\, \mathbf{H}_{iron}. \tag{9}$$

Recognizing that the first term on the right represents nothing more than the flux density in the iron alone, the full prescription for material curve scaling is thus

$$\begin{bmatrix} H \\ B \end{bmatrix}_F = \begin{bmatrix} 1 & 0 \\ \mu_0(1 - s) & s \end{bmatrix} \begin{bmatrix} H \\ B \end{bmatrix}_{iron}. \tag{10}$$

If this scaling is applied to every point of the B-H curve given for the stator iron, the corresponding curve for the fictitious composite material is obtained.

It should be noted carefully that any analysis performed on the fictitious machine built with material F cannot produce correct flux densities in the *iron*, only in material F. To determine the local saturation level of the stator iron, it is necessary to invert equation (10),

$$\begin{bmatrix} H \\ B \end{bmatrix}_{iron} = \begin{bmatrix} 1 & 0 \\ -\mu_0\,(1 - s)/s & 1/s \end{bmatrix} \begin{bmatrix} H \\ B \end{bmatrix}_F, \tag{11}$$

and to apply this inverse transformation to the flux densities computed for the fictitious material.

An alternative, and in some respects simpler, approach to the ventilating duct problem is to refer all calculations to the rotor. Proceeding in this way, the axial length of the machine is taken as the length of the iron actually present in the stator, that is, its true axial length multiplied by the fill factor s, as defined above. The stator magnetization curve is then used as it stands, but the rotor curve is modified, scaling it so as to create a new fictitious material G. For a given magnetic field $\mathbf{H}$, the new material must be more permeable than the true rotor iron R; in contrast to a scaling referred to the stator, there is no paralleling air path and the scaling required is very simply

$$\begin{bmatrix} H \\ B \end{bmatrix}_G = \begin{bmatrix} 1 & 0 \\ 0 & L_{stator}/L_{rotor} \end{bmatrix} \begin{bmatrix} H \\ B \end{bmatrix}_{iron}. \tag{12}$$

This transformation is easy to compute and easy to invert. Of course, it will yield the correct flux densities in the stator but not in the rotor, in contrast to the stator-based scaling of equations (10)–(11).

Every magnetization curve scaling technique can be made to compute one quantity of interest correctly, but all other quantities may then require inverse scaling. Scaling as in equation (10) will produce the correct total stator flux, and hence will compute excitation requirements and terminal voltages correctly; but local flux densities in the stator (not the rotor!) will require inverse scaling. Conversely, scaling as in (12) will get the stator flux densities right, but not the rotor values. Which one is best in a particular application will of course be a matter of taste. In the absence of any other basis for choice, it is advisable to select that which will involve the fewest transformations, just to keep work simple.

Packet Fringing

The scaling techniques discussed in the foregoing furnish first-order corrections. For practical cases, either one is often sufficient. However, it fails to account for the fact that axial fringing flux must exist near the packet ends in the ventilating ducts, and it gives no indication of how large or small this axial fringing flux might be. Such fluxes must exist because flux lines that cross the rotor surface at a point opposite a ventilating duct need to take a partly axial, partly radial path in order to enter a stator iron packet.

A stator packet, half of the ventilating duct to either side of it, and the rotor iron across from it are shown in Fig. 19. The axial direction of the machine corresponds to the horizontal direction in the figure. Although the ventilating ducts shown are relatively narrow compared to the machine air gap, some longitudinal flux may be seen entering the stator packet at its ends. This axial flux component may cause some eddy current loss, but it is not usually considered serious because this local loss occurs precisely at the ventilating duct surface, where cooling is at its best.

On examining Fig. 19, the radial (vertical) flux density at the rotor surface is seen to vary along the rotor surface, being largest at the packet centerline and smallest at the duct centerline, just as one might expect. Perhaps less obvious is the behavior of the radial flux component at the stator surface: it must rise near the packet ends, in order to accommodate the flux which crosses the rotor surface near the ducts. This rise amounts to perhaps 5% above the average value in the case investigated here, a percentage of no great consequence in an unsaturated machine but one likely to require noticeable additional excitation if the machine iron is run well into saturation.

A suitable way of correcting for axial fringing is to make yet another adjustment to the material properties. The correction technique is quite similar to that alluded to in connection with the butt-gap problem in transformer laminations, and involves the following steps.

1. The fringing field problem is solved, as in Fig. 19, for a set of excitation levels appropriate to the rotor and stator materials.
2. The stored energy is evaluated for a region consisting of the air gap and a portion of the stator packet as wide as the air gap.

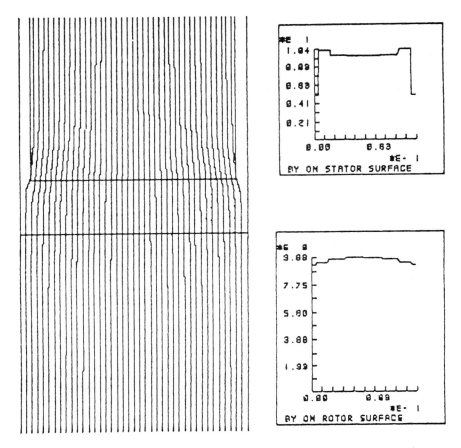

Figure 19. Stator packet between two ventilating ducts, and its associated rotor iron. The drawing horizontal axis corresponds to the axial direction in the machine.

3. For each excitation level, the average flux density in the rotor (and air gap) is calculated and the energy stored in the air gap is computed, assuming that the flux density in the air gap is uniform.

4. The hypothetical air gap energy is subtracted from the known stored energy of air gap and nearby stator iron, to yield an energy difference.

5. The difference in energy is accounted for by filling the space next to the air gap, and one air-gap width broad, with a fictitious material instead of stator iron. This fictitious iron fills both packet and duct space. Its magnetization curve is computed from the energy known to be stored in it for various known flux densities.

In summary: the inhomogeneous stator is replaced by two fictitious materials. The space occupied by stator iron and ventilating ducts is filled with a material whose magnetization curve is given by equation (10), with the exception of the tooth-tops, i.e., the region immediately adjacent to the air gap. There, the true materials are replaced by a fictitious material whose *B-H* curve is computed to yield the right stored energy when packet fringing is taken into account.

End Effects

A great many magnetic devices, including in particular recording heads and electric machines, can be analyzed on a substantially two-dimensional basis, but the three-dimensional fields at the device ends are not altogether negligible. However, there are many instances in which the field behavior in the end-region can be approximated tolerably well in a separate analysis; end effects can then be incorporated into a two-dimensional model as corrections. In this case, the treatment of end effects forms a particular case of the subproblem approach.

In Fig. 20 a rough diagrammatic view is shown of an electric machine end-zone. To permit approximate two-dimensional analysis, it is assumed that the end-zone flux has no azimuthal component, but possesses only radial and axial components. Furthermore, the rotor and stator iron are taken to be infinitely permeable. The analysis is therefore confined to a roughly L-shaped region bounded by the rotor and stator surfaces, which the flux lines are required to meet orthogonally. Two flux-line boundaries must be established to make the problem solvable. It is reasonable to

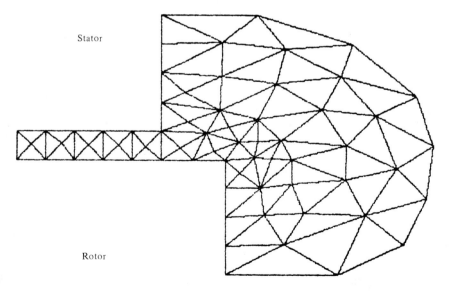

Figure 20. End zone of a short-stator electric machine. Leakage flux path reluctance is estimated from an approximate two-dimensional boundary-value problem.

assume that flux lines must cross the air gap orthogonally at some moderate distance (perhaps three air gap widths) from the end, so a flux line is assumed to cross there. Some distance from the gap end, the end-region model is terminated by assuming another flux-line boundary to exist. There is usually no solid basis on which to place such a flux-line boundary. But for most purposes its precise location is probably not very important, because the flux density, stored energy density, and indeed all other electromagnetic variables fall off quite rapidly with distance from the gap-end. The resulting boundary-value problem is linear and simple. As indicated in Fig. 20, it comprises the end-region air space. The boundary conditions are of homogeneous Neumann type at the iron surfaces and of Dirichlet type along the flux lines. The latter are inhomogeneous, so that the total flux crossing from rotor to stator is prescribed.

The end-leakage flux path reluctance may be estimated from the approximate problem as described. The flux density at the left edge of Fig. 20 (in the air gap proper) is multiplied by the length of stator actually represented in this analysis, to obtain the air-gap flux that would exist were the air-gap flux density to remain invariant right up to the end of the stator. This amount is subtracted from the prescribed total flux in the end-zone model, to yield the end leakage flux. The magnetomotive force between rotor and stator is next determined, and divided by the end leakage flux; the result is the estimated reluctance of the end leakage paths.

Where the total stored energy in a magnetic device is of primary importance, the additional flux paths provided by the end-zone merely increase the effective machine length. It suffices to determine the ratio of end-zone reluctance to the reluctance of a unit length of air gap (better still, a unit length of the machine magnetic path) and to increase the machine length for computation purposes proportionately. While this approach is relatively crude, it is highly effective wherever the end-zone contribution is small.

In some electric machine analyses, a key issue is the correct representation of mutual and leakage fluxes, that is to say, the identification of those parts of the machine flux which do or do not link both the rotor and stator conductors. Purely two-dimensional analysis based on an effective length increase, as described above, is then inappropriate, because increasing the effective length increases the mutual flux but does not introduce additional leakage flux. One practical way for increasing the leakage in two-dimensional analysis, so as to bring it to the right total level, is to increase the interpolar leakage slightly. For example, two-dimensional analysis of the dc machine of Fig. 14 ignores the end leakage flux. To include the right amount of leakage flux in a machine performance analysis, the air space next to the field poles, and possibly the field winding material, may be assigned a relative permeability a little higher than unity. In effect, the missing axial leakage flux is then replaced by inclusion of some extra leakage flux in the azimuthal direction.

Planning the Solution

It should be evident from the examples given above that the physical formulation of a problem depends greatly on just what results are desired. Even though only one physical device exists in reality, it may be represented by many different idealized models, and some of these may emphasize one aspect of the problem over another. Equally, the computational formulation of a problem, for example selecting and grading the mesh, is of importance to the user. Getting the mathematical physics of the problem right is essential to getting the answers right; choosing the right computational approach is important for good accuracy and computing time. Setting up a problem for computation is much like writing a computer program: it should be carefully planned before even approaching the computer. Much time and effort can otherwise be wasted in abortive attempts.

Objectives

The analyst generally does well to imagine himself in the role of an experimenter about to make a set of measurements on the device to be analyzed. As for the experimenter, so for the analyst it is essential before starting out to settle the most important question: *what data should this experiment yield?* Models useful for determining critical local flux densities are not necessarily the most efficient models for finding terminal impedances, and the best efforts of men as well as machines can be uselessly dissipated by working with wrong or excessively detailed models. It is generally wise to work with fairly minimal models, refining them gradually as needs become apparent. It is not wise to launch straightaway into construction of a physical model of great computational and mathematical complexity, in the hope that it can be made sufficiently detailed to be good enough for everything. In fact, if there is a common fault among beginning users of CAD tools, it is probably a tendency to overkill in mesh modelling.

Modelling choices and techniques are generally determined by objectives. The most usual ones in magnetics are:

1. Determination of specific terminal characteristics (impedances, terminal voltages, power losses, stored energies).
2. Determination of specific local flux densities (e.g., density in the recording medium below a recording head).

The objectives to be pursued are critical in determining the physical approximations to be employed. Commonly, these may involve decisions on

1. whether a two-dimensional (sectional) analysis is sufficient, or whether some form of end-effect compensation should be included,

2. whether materials can be assumed to be linear, or whether nonlinearity forms an essential part of the problem.

If time-harmonic phenomena (alternating current excitations) are to be treated, or if circuit parameters such as inductances are to be found, then very often an assumption of linearity is essential even in devices containing iron. The very notion of impedance is undefined in nonlinear circuits!

Mesh Modelling

Meshes suited for determining local events generally have different characteristics from meshes suited to the computation of global characteristics of a device. If both are required, it is often a good idea to produce a single, quite coarse, mesh first. This coarse mesh can then be used as a basis for refinement in several directions for the several and various requirements to be satisfied.

It seems almost redundant to point out that for determination of local flux densities, local mesh refinement is required. In Fig. 11(b), for example, the mesh is locally refined to allow determination of the crucially important flux density below the recording head. In other parts of the device, the mesh shown probably represents overkill. There appears to be no particular reason for equally fine meshing of the entire iron member, other than perhaps ease of mesh generation.

In contrast to finding local details, terminal parameter calculations usually require a fineness of meshing which very roughly parallels the energy storage density in the device. As a simple rule of thumb, some 200 - 600 elements usually suffice to find the terminal parameters of simple electric machines of standard design, to the accuracy justified by assuming that end effects do not matter. For example, the meshes shown for the turboalternator of Fig. 13 and the dc machine of Fig. 14 involve a few hundred triangular elements per pole pitch. They can be expected to predict most of the usual performance criteria (e.g., excitation curves) to within two or three percent, provided some rough compensation is made for end effects.

Advance Planning

Most analysts find it easiest to plan a problem by taking time to prepare a clear and detailed sketch of the device, then a sketch of the boundary-value problem to be solved. The latter should include not only the coordinates of each fiducial (key) point required to define the model but also any relevant notes about the units and the coordinate system employed. Regions sharing common properties should be clearly indicated. It should be noted that current densities in go-and-return windings have opposite senses, so that region or material identifications for, say, the two sides of a single winding should be distinct.

It is wise to allow for any future mesh refinement or alteration at the outset. For example, the mesh used in the accelerator magnet analysis of Fig. 16 is based on three original sketches, one for each of the various bolthole placements. The mesh includes provision for all three, thus permits easy modification when required.

Postprocessing Operations in CAD

The ensemble of activities by which the engineering significance of a mathematical field solution is evaluated is generally termed *postprocessing*. It includes the derivation of specific numerical results as well as their graphical presentation. A postprocessor is a major part of any design system since it allows relevant data to be extracted from the solution and presented in a way that has meaning to the user. Postprocessing should desirably be an interactive process, allowing the designer to query the solution. This chapter outlines the major requirements of postprocessing, while the following one describes a particular postprocessor structure by way of illustration.

Postprocessing is the activity of converting mathematical solutions into engineering results. This chapter examines the operations required in a selection of postprocessing tasks, in an attempt to exhibit unity in the processing requirements that underlie the great diversity of applicational needs. Thus the discussion here will begin by examining simple but illustrative design problems, and will then generalize to broader issues applicable not only to the examples treated but also to other cases.

Inductance Calculations

The calculation of terminal inductance values is probably the most common single requirement in magnetic device design. Despite its frequent occurrence, the determination of inductance is fraught with subtle difficulties not always evident at first glance. Examining a simple magnetic-core reactor and considering the methods available for calculating its inductance is therefore both indicative of general methods and useful in its own right.

A Simple Inductor

A simple, two-dimensional inductor may be formed by winding a coil around a highly permeable core, as shown in Fig. 1. It is assumed, for the

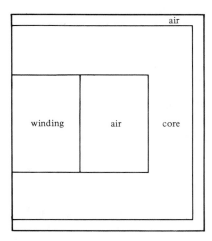

Figure 1. A simple inductor, assumed to extend infinitely in the direction normal to the paper.

sake of simplicity, that the device extends sufficiently far into the paper to allow purely two- dimensional analysis. Not only is the finite element analysis of the device made easy by this assumption, but basic conventional design rules may be applied for comparison when the inductor is simplified in this way.

Since this device is to furnish a prescribed value of terminal inductance, that is, to store energy in the magnetic field in its core, certain further simplifying assumptions are possible in the analysis. In addition to taking the field problem to be two-dimensional, two main simplifications will be made: the core iron will be assumed lossless, and external leakage flux, i.e., leakage flux not in the window area of the core will be neglected.

Assuming iron loss due to eddy currents to be absent permits finding the magnetic field by means of one or more static field solutions. The only loss accounted for will then be the ohmic loss in the winding conductor itself. This loss cannot be calculated from a magnetic field solution, but must be computed from the known wire resistance. The inductor will therefore be represented, to an approximation adequate for most purposes, by the equivalent circuit shown in Fig. 2.

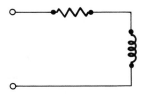

Figure 2. An equivalent circuit of the simple inductor shown in Fig. 1, assuming lossless iron.

The principal part of all the flux which links the current-carrying winding will very likely be contained in the iron core, while some small portion will close through the air space within the window opening of the core. Only a very minor amount will take a path outside the core itself. Hence, little error will be incurred in not modelling the air space and winding outside the core; it suffices to take the outer edge of the core perimeter to be a flux line and to model only the iron, air, and winding inside this flux-line boundary. Furthermore, the symmetry of the core and winding make it necessary for the core centerline to be a line of symmetry, and hence the separatrix flux line which separates the flux lines of clockwise circulation from those which close in a counterclockwise sense. It is therefore necessary to model only one half, say the right half, of the inductor. A finite element model of this half-problem appears in Fig. 3. While moderately crude, this discretization probably suffices to produce a magnetic field solution adequate for inductance computations. Any lingering doubts about the amount of external leakage flux may of course be dispelled by employing a similar model, but with some of the exterior air space explicitly included.

Physical realization of such a device is invariably subject to many requirements other than electromagnetic, such as weight limitations or the cost of material used. These constraints not infrequently will be handled

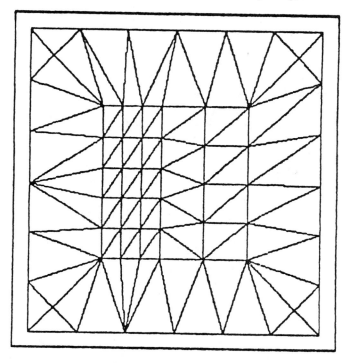

Figure 3. Finite element model used for analysis of the simple inductor.

by other CAD or CAE systems. Only its electrical design will be considered here.

Definitions of Inductance

The computation of inductance is a simple affair in principle: numerical values taken from a field solution are substituted into an appropriate expression that gives the inductance in terms of the magnetic field, and the result is evaluated. Unfortunately, there is no single definition of just what constitutes inductance for a nonlinear inductor. Indeed, there are several accepted definitions, of which two may be considered fundamental:

1. The number of flux linkages of the winding, divided by the current in the winding.
2. The energy stored in the inductor, divided by one-half the current squared.

All other accepted definitions make reference to either the flux linkages or the stored energy; they may therefore be considered variants of the above. Both definitions give identical results for linear inductors. In the nonlinear case, however, they do not. Which definition should be used, if indeed either, then depends very much on the uses to which the result will be put—or, what amounts to the same thing, the kind of measurement which the analysis is intended to mimic. Because periodic excitations (not necessarily sinusoidal) are of frequent interest, the second definition is often further refined in one direction or another, for example, to refer to the *average* stored energy and the *fundamental component* of current.

There are two experimental ways of measuring inductance, which parallel the two analytic definitions. One seeks to determine flux linkages more or less directly, the other measures time-averaged energy storage.

Direct measurement of flux linkages in an inductor is readily accomplished by a method widely used since the invention of the galvanometer and generally called the *ballistic fluxmeter* method. The principle is to open- circuit the inductor winding suddenly and to integrate the resulting terminal voltage over time so as to determine the total flux linkages that existed prior to open-circuiting. If $n\phi$ is the number of flux linkages, Faraday's law prescribes an open-circuit terminal voltage

$$e = - \frac{d\,(n\phi)}{dt}\,, \tag{1}$$

whose time integral, as measured by an integrator circuit, is equal to the initial value of flux linkages:

$$n\phi(0) = - \int_{-\infty}^{0} e(t)\,dt. \tag{2}$$

In practice, the current is usually reversed rather than merely disconnected. The number of flux linkages is thereby doubled, reducing the
error of measurement slightly and simplifying the experimental procedure; but more importantly, hysteretic effects are substantially reduced
or even eliminated. This method is directly related to the first, flux linkage
based, definition of inductance. Its noteworthy aspect is that the measured
value is not directly dependent on the shape of the material B-H curve
and is therefore not related to the amount of energy stored; various
different B-H curves could yield the same value of flux linkages. In other
words, the measured result is truly the total of flux linkages; no approximations are involved. Inductance is then determined as the ratio

$$L = \frac{n\phi(0)}{i(0)}. \tag{3}$$

A second, very common, technique for measuring inductance is to find the
root-mean-square values of voltage and current when a time-sinusoidal
excitation is applied to the device. The inductance value is then found as
the measured reactance divided by angular frequency,

$$L = \frac{X}{\omega} = \frac{E_{rms}}{\omega I_{rms}}. \tag{4}$$

This measurement is closely related to the second or *energy* definition of
inductance, since the reactance is in principle the amount of energy stored
per unit of coil current, averaged over the ac cycle. Let it be supposed for
the moment that a source of sinusoidal current is connected to the terminals of the inductor, Fig. 2,

$$i(t) = I \sin \omega t. \tag{5}$$

The corresponding flux linkages will reflect the shape of the B-H curve of
the core material, so the flux linkages, and hence the terminal voltage, will
not be sinusoidal but must contain all odd harmonics as well:

$$e(t) = \sum_{k=1}^{\infty} E_{2k-1} \cos\left[(2k-1)\omega t + \theta_{2k-1}\right]. \tag{6}$$

When root-mean-square measurements are made, the measured voltage
will include a fundamental term and an additional contribution from
every harmonic term. That is to say,

$$E_{rms} = \frac{1}{2} \sum_{k=1}^{\infty} E_{2k-1}^2 \tag{7}$$

so that equation (4) reads

$$L = \frac{1}{2\omega I_{rms}} \sum_{k=1}^{\infty} E_{2k-1}^2. \tag{8}$$

The laboratory measurements implied by (8) are easily carried out, so that (8), as well as a similar equation based on sinusoidal applied voltage and a nonsinusoidal resulting current, are commonly used as inductance definitions. To calculate inductance by simulating this measurement is feasible, but computationally a bit lengthy. Strictly speaking, it will be necessary to solve a static field problem for every one of a set of current values, from zero up to the peak value; then to find the corresponding flux linkages, and to differentiate these so as to obtain the voltage waveform. From the voltage waveform, the root-mean-square value can then be computed. Such a computation is only very rarely carried out, not only because it is complicated but also because the inductance value as defined by (8) is not ideally suited to many applications. Indeed, quite a few practical needs are better satisfied by a similar definition, but referring to the fundamental components only,

$$L = \frac{E_1^2}{2\omega I_{rms}}.$$ (9)

This definition is particularly useful if the winding under consideration is one of several, and it is known in advance that any reasonable interconnection of the windings will result in a great deal of harmonic cancellation. Such interconnections are often encountered in electric machines.

It is possible to make measurements which seek to determine the stored energy corresponding to a specific instantaneous current value, so as to apply directly the energy-based analytic definition of inductance. The principle is exactly opposite to that of the ballistic fluxmeter experiment: the inductor coil is suddenly connected to a pure resistance, and the instantaneous power in the resistor is integrated over time:

$$W = \int_0^\infty \frac{\partial}{\partial t} [n\phi(t)] \, i(t) \, dt.$$ (10)

As the energy stored in the inductor is gradually dissipated in the resistor, the current $i(t)$ falls, and the flux linkage $n\phi(t)$ falls with it in accordance with the shape of the B-H curve of the core material. The inductance value thus measured corresponds truly to the second definition given above, and faithfully reflects the saturation characteristics of the core material. Such measurements, however, are rarely made because they are experimentally difficult to carry out.

Inductance from Flux Linkage

When a CAD system is used to solve the magnetic field in the inductor described above, a static solution is normally produced. Static solutions can be closely related to the first, flux-linkage-based, definition of inductance, and the corresponding calculations are quite easily carried out.

Let the magnetic field be determined in the simple inductor described above. To make the matter concrete, the core is taken to have external

dimensions 0.8 m by 0.4 m, so that the half modelled in Fig. 3 measures
0.4 m by 0.4 m; the window size is 0.2 m by 0.2 m. The coil occupies a
space 0.08 m by 0.2 m. The *B-H* curve for the model is as shown in Fig. 4.
It may be noted that the curve is carried up to quite high saturation levels; the current increments along the *H* axis are 10000 ampere-meters
between ticks, while the values of *B* extend up to 2.48 tesla. A solution,
showing flux lines in the core and in the window space of the inductor,
appears in Fig. 5. The coil excitation in this case is 500 ampere-turns.
With the mean flux path length in the magnetic core of the order of one
meter, it is clear from Fig. 4 that this solution is essentially a linear one;
the magnetic material is working well within the linear region of its
saturation curve throughout. The material permeability is correspondingly
high and, as might be expected, practically all the magnetic flux is
confined to the iron core.

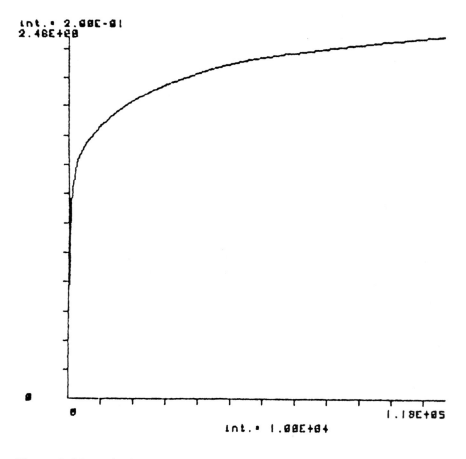

Figure 4. Magnetization curve used in inductance calculations for the simple
inductor.

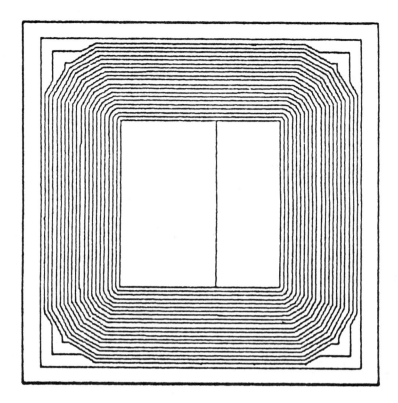

Figure 5. Flux distribution in the simple inductor at low saturation level (coil excitation 500 ampere-turns).

Inductance calculations based on flux linkage counts are very simple to carry out if the flux is principally confined to a clearly defined iron member or path, as it is in Fig. 5. In effect, it suffices to pretend that the winding can be replaced by a single filamentary coil and to calculate the flux spanned by that coil. In the case shown here, it does not even very much matter where the filamentary coil is placed, provided only that one side of it lies somewhere within the core window, while the other side is placed at a corresponding point in the other window. The precise position is unimportant, for so long as the coil is threaded by the iron core, it will link practically the same amount of flux.

Suppose next that a hypothetical single-turn filamentary coil is placed in the window space of the inductor as shown in Fig. 6(a). The flux ϕ linked by this turn can be expressed in terms of the flux density B and the area enclosed by the coil as

$$\phi = \int \mathbf{B} \cdot d\mathbf{S}. \tag{11}$$

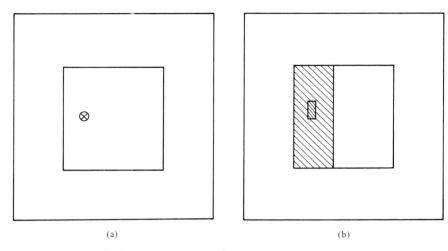

(a)   (b)

Figure 6. (a) A single-turn conductor or thin-sided coil in the inductor structure. (b) A slim-sided coil as part of a thick winding.

Substituting the vector potential **A** for the flux density **B**, in accordance with

$$\operatorname{curl} \mathbf{A} = \mathbf{B}, \tag{12}$$

one obtains

$$\phi = \oint \mathbf{A} \cdot \mathbf{dl}, \tag{13}$$

where the integration is around the closed contour formed by the coil, i.e., following the wire of which the coil is made. Because the problem is essentially two-dimensional, the end contributions to the integral in (13) may be ignored. Along the two long sides of the coil, which lie within the window space of the core, the vector potential has constant values, say A_L on the left and A_R on the right. If the length of such a side is Z, then the integral in (13) takes the simple form

$$\phi = Z(A_R - A_L). \tag{14}$$

For a thin multi-turn coil, a similar development may be carried out as for the single turn above. If the coil is made of n turns, then integration must be carried out over the spiral contour traced by the wire. If the turns are tightly packed and occupy little space, the same argument may be applied to each one; the flux linkages $n\phi$ are therefore n times larger,

$$n\phi = n Z(A_R - A_L). \tag{15}$$

To keep the total exciting ampere-turns I of the n-turn coil the same as those of the single turn, the current must be reduced by a factor of n, to

$i = I/n$, because the same current is threaded through the core window n times. The inductance is therefore finally calculated as the number of flux linkages no divided by the coil current I/n:

$$L = \frac{n\phi}{I/n} = \frac{n^2 Z (A_R - A_L)}{I}. \tag{16}$$

It will be noted that the inductance value is proportional to the square of the number n of turns in the winding. Computing the inductance of a multi-turn coil with thin sides is thereby reduced to determining the vector potentials A_R and A_L at its two sides and multiplying by $n^2 Z$, mathematically straightforward operations.

The very simple flux linkage method of calculating inductance is not applicable to windings that fill an extended portion of space, because there is no particular point at which to measure the vector potential. In effect, thickening the coil sides requires forming some sort of average value of the vector potential over the space occupied by the winding. The mathematical development in this case is fortunately still quite simple, as may be seen from the following.

A thick winding may be regarded as a set of individual turns, or perhaps as a set of smaller, thin, multi-turn coils, all connected in series. The total flux linkage of the thick winding must then be the sum of the individual flux linkages of its component parts. In Fig. 6(b) the simple inductor appears again, with a particular small group of adjacent turns highlighted; if this group has n_k turns, say, then its flux linkages are

$$(n\phi)_k = n_k (A_{kR} - A_{kL}), \tag{17}$$

where it is assumed that the group is compact enough for the potential values at its left and right sides, A_{kL} and A_{kR}, to be substantially constant over the extent of the group. Summing over all the m elementary coils that make up the winding, the total flux linkage is

$$n\phi = \sum_{k=1}^{m} (n\phi)_k = \sum_{k=1}^{2m} n_k \frac{J}{|J|} A_k. \tag{18}$$

Here the last summation is carried out over coil *sides*, rather than coils, in the interest of generality. To ensure that oppositely directed sides enter into the summation with opposite signs, each term has attached to it the local direction of the current density **J**.

The flux linkage calculation as set out in equation (18) requires the winding to be divided into sections and their individual contributions to be summed. The inconvenience of explicit subdivision is avoided by using a more convenient formulation of the problem. To reformulate, it suffices to note that, since all turns in the winding are in series and therefore carry the same current, the *signed* number of turns of a particular coil side, as in the rightmost member of (18), may be written

$$n_k \frac{J}{|J|} = \frac{\int_{S_k} J \, dS}{ni} \, n, \tag{19}$$

where the the surface of integration S_k is the cross-sectional area of the kth coil side. The numerator in equation (19) therefore represents the number of ampere-turns contributed by the kth elementary coil side, while the denominator equals the total number of ampere-turns for the whole coil. In simple words, (19) thus says that the number n_k of turns in the elementary coil side can be found by determining what fraction of total ampere-turns it contributes. Substituting (19) into (18), there results

$$n\phi = \frac{1}{i} \int_S A \, J \, dS, \tag{20}$$

where i represents, as previously, the current in the coil. The winding inductance is again calculated from its definition as the number of flux linkages per ampere:

$$L = \frac{Z}{i^2} \int_S A \, J \, dS, \tag{21}$$

where Z represents, as before, the length of the winding in the z direction. While this inductance expression is valid for two-dimensional cases as given here, it is readily generalized to cover situations where the coil sides are not straight, indeed to coil configurations not describable by two-dimensional approximations at all. Omitting the mathematical details, the result turns out to be

$$L = \frac{1}{i^2} \int_U \mathbf{A} \cdot \mathbf{J} \, dU, \tag{22}$$

where U is the volume occupied by the winding.

Although the number of turns n does not appear explicitly in equation (22), L is dependent on it through the values of J and A. Suppose, for example, that the number of turns is increased from n to N, and that the current i is correspondingly reduced to $(n/N)i$. The total number of ampere-turns remains unchanged in this process; so does the current density J. The vector potential A, however, depends only on J, and therefore remains unchanged as well. Thus, the numerator in (22) is not changed, while the denominator is altered by $(n/N)^2$. The inductance therefore increases by the factor $(N/n)^2$, in accordance with equation (22).

Stored Energy and Inductance

The inductance value of a reactor may be defined in terms of the magnetic energy W stored in it. This definition relies on the familiar energy expression

$$W = \frac{1}{2} L\, i^2,\qquad (23)$$

where i is the terminal current of the inductor, as in the above. If the stored energy W can be determined, then the inductance value follows,

$$L = \frac{2W}{i^2}.\qquad (24)$$

The stored energy contained within a magnetic device may be found by integrating the stored magnetic energy density w over the volume U of the device:

$$W = \int_U w\, dU.\qquad (25)$$

But at any point in a magnetic material, the stored energy density is given by the area to the left of its magnetization characteristic:

$$w = \int_0^B \mathbf{H}(\mathbf{b}) \cdot \mathbf{db},\qquad (26)$$

where $\mathbf{b}$ is a dummy variable which follows the flux density along the B-H characteristic, which in general is a vector relationship, up to its final value B. The inductance of the inductor may therefore be calculated by using the relationship obtained by combining equations (24)–(26):

$$L = \frac{2}{i^2} \int_U \int_0^B \mathbf{H} \cdot \mathbf{db}\, dU\qquad (27)$$

Evaluation of this quantity is fairly straightforward in well designed CAD systems, which have access to the material B-H curves. After all, the magnetization curve of every material must be known to the system if solution of the field problem is to be possible in the first place.

In several older magnetics CAD systems, the material magnetization characteristics are not actually available at postprocessing time; instead, the material reluctivity (inverse permeability, i.e., the value of H/B at the solution point) is known. In such circumstances, a rough approximation based on linear theory can sometimes lead to useful results. Linear or not, equation (26) may always be rewritten as

$$w = \int_0^B \nu\, \mathbf{b} \cdot \mathbf{db}.\qquad (28)$$

For a magnetically linear material, the reluctivity may be moved across the integral sign, so that (28) becomes

$$w = \nu \int_0^B \mathbf{b} \cdot \mathbf{db} = \frac{B^2}{2\mu}. \tag{29}$$

Although this expression is not correct for nonlinear materials, it sometimes leads to results of adequate accuracy. That is to say,

$$L_{approx} = \frac{2}{i^2} \int_U \frac{B^2}{2\mu} \, dU \tag{30}$$

can furnish useful approximations where better ones are not available. Although the substitution of linear for nonlinear magnetization characteristics may seem a very crude approximation, it is frequently permissible because the use of *inductance* as a circuit parameter is often confined to near-linear cases anyway. If the approximation is very bad, then quite likely the use of inductance for whatever further purpose is likely to be a bad idea also!

An Example Calculation

The methods of inductance calculation discussed above may be illustrated by giving numerical examples. These relate to the inductor of Fig. 1, whose core material is characterized by the curve of Fig. 4.

A field solution obtained for very low excitations is shown, and briefly discussed, in Fig. 5. The flux density in that solution is everywhere fairly low, as the result of the low excitation value; the material permeability is high, and the flux is therefore almost entirely confined to the iron core. A different situation obtains when the excitation is raised to a much higher value. A second solution, computed for an excitation of 100 kiloampere-turns, is shown in Fig. 7. It is clear from Fig. 7 that most of the flux still resides in the iron core; but the leakage flux crossing the core window is considerably larger than in the earlier case. Even at this high saturation level, however, the leakage flux still represents only a modest fraction of the total flux, so that the same methods of inductance calculation may be applied, as is indeed done below.

Numerical values of flux densities at a few selected points are shown for both solutions in Table 1. The increase in leakage flux is immediately evi-

Table 1. Flux density values at selected points

Excitation (kA-turn)	Left limb T	Upper limb T	Right limb T	Lower limb T	Window mT	Winding mT
0.5	0.99	0.99	0.99	0.99	0.10	0.50
100	2.57	2.45	2.35	2.45	140.	13.0

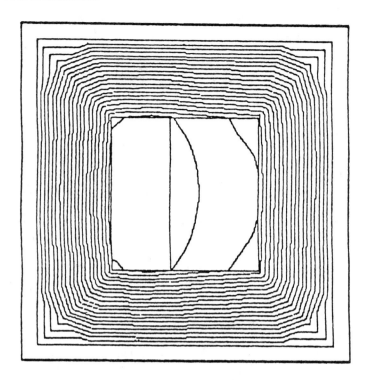

Figure 7. Flux plot for the simple inductor of Fig. 1, with a coil current to produce 100,000 ampere-turns.

dent. Flux densities in the iron portions of the inductor rise by a factor of about 2.5 from one solution to the other; but the flux density in the winding space rises by 26 times, while that in the window air space increases to 1400 times its value at low saturation levels!

Inductance values may be calculated from either solution, using the several techniques given above. Four values of inductance are given in Table 2 for either case: two computed using the flux linkage approach, two using the energy approach. The flux-linkage-based calculations include the approximate simple one,

$$L = \frac{n\phi}{I/n} = \frac{n^2 \, Z \, (A_r - A_L)}{I}, \qquad (16)$$

as well as the more precise

$$L = \frac{\int_U \mathbf{A} \cdot \mathbf{J} \, dU}{i^2}. \qquad (22)$$

The energy-based values are computed using first the full and correct stored energy expression,

$$L = \frac{2}{i^2} \int_U \int_0^B \mathbf{H} \cdot \mathbf{db} \; dU \qquad (27)$$

and then also the quasi-linearized approximation

$$L = \frac{2}{i^2} \int \frac{B^2}{2\mu} \; dU \qquad (30)$$

Results obtained by these four methods are shown in Table 2. They are expressed in microhenries, for an inductor with a winding assumed to contain exactly one massive turn; for realistic windings, the inductance values should be multiplied by n^2. For example, the correct inductance values for a 1000-turn winding are those shown in Table 2, multiplied by 10^6, that is, with Table 2 read as containing values in henries instead of microhenries. It is immediately evident that the two saturation levels are indeed extreme, their corresponding inductance values varying by about two orders of magnitude.

Table 2 is so arranged that the two most accurate values, those computed by using equations (21)–(22) and (27), occupy the two columns in the middle of the table. As can be seen from Table 2, the values are in very good agreement with each other for the unsaturated case.

The approximate energy-based calculation of equation (30) is seen to yield results in close agreement with those based on flux linkage totals. This phenomenon is perhaps not so surprising after all if one notes that the approximate energy calculation employs only the static permeability B/H, in other words, that the calculation relies heavily on one or a few points on the B-H curve; and so do the flux linkage methods. In essence, all three replace the B-H curve with a straight line that connects the origin with the current working point. The true nonlinear energy-based value, on the other hand, takes the curvature of the B-H characteristic fully into account.

Given that the different methods of inductance calculation can yield results as far apart as 0.974 and 2.556 for the same physical situation, which is correct? The answer, of course, is: both! It is important to keep in mind that the difference lies not so much in the computation methods

Table 2. Computed inductance values

Excitation (kA-turn)	L via flux linkages (microhenries)		L via stored energy (microhenries)	
kA-turn	(16)	(21)	(27)	(30)
0.5	198.96	198.96	181.50	198.96
100	2.556	2.556	0.974	2.556

as in the definitions of *inductance*. In fact, two different quantities are calculated and listed in Table 2, since the various definitions of what constitutes inductance coincide for the linear case—as indeed they do for the low-saturation case in Table 2—but are quite different for strongly saturated materials.

In practical design problems, difficulties are fortunately not as bad as they may seem from the above discussion. Most inductor designs have an air gap in the magnetic circuit to control the inductance under varying operating conditions. The inductor shown in Fig. 7 has been modified in Fig. 8 to include an air gap in one limb. The model also has to be changed to include some air space in and around the gap, in order to allow for the fringing flux which can now appear around the air gap. The gap introduced is 10 cm wide. The solution was recomputed for the two current levels described earlier and Fig. 8 shows the flux distribution for the saturated, 100 kA, case. The inductance values for the two current levels are shown in Table 3.

As can be seen, the variation in inductance between the saturated and unsaturated conditions is now considerably less, being of the order of 15% rather than the 80% without the air gap. The reason for the change is that

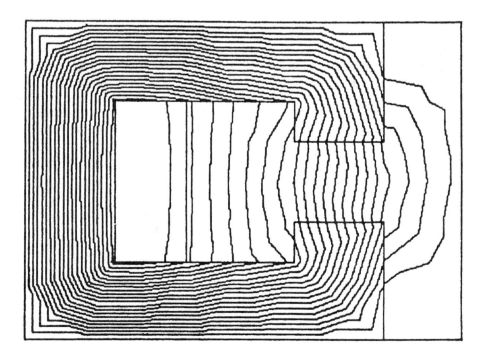

Figure 8. The inductor with air gap, subjected to an excitation of 100 kiloampere-turns.

Table 3. Computed inductance values with air gap

Excitation	L via flux linkages (microhenries)	L via stored energy (microhenries)	
kA-turn	(21)	(27)	(30)
0.5	2.750	2.750	2.750
100	2.177	1.823	2.177

the air gap dominates the problem in terms of where the energy is stored. The results detailed in Tables 1–3 were calculated on a finite element mesh having 76 nodes and 122 elements (for the problem in Fig. 8). The results of Table 3 were recomputed for a model having 543 nodes and 1039 elements but describing the same physical device. Considerable care was taken in the placement of the elements in the refined mesh in order to get a good solution for the flux distribution in the core and air gap of the inductor. The inductance values, corresponding to those in Table 3, for this highly refined model are shown in Table 4.

The results for the fine mesh give inductance values which vary by about 6% on average from those calculated with the coarse discretization. If a similar test is performed without the air gap present, the variation is about 1%. These results highlight the fact that the discretization chosen for a particular problem depends heavily on the results which are to be obtained—the refined mesh took approximately 20 times longer to solve than the coarse system and yet the inductance values, based on global energy storage calculations, vary by less than 6%. However, if the flux distribution is considered and the degree of saturation at various parts of the circuit is of prime importance, then the refined mesh produces considerably more accurate results because the coarse system does not allow any flux redistribution around the magnetic circuit.

A Transformer Design Problem

In the design of small transformers, it is important to determine the conventional transformer equivalent circuit parameters. Not only are they significant as evaluation criteria, but they often appear as key points in

Table 4. Computed inductance values with air gap—refined mesh

Excitation	L via flux linkages (microhenries)	L via stored energy (microhenries)	
kA-turn	(21)	(27)	(30)
0.5	2.938	2.938	2.938
100	2.237	1.820	2.358

customer specifications. These parameters are experimentally determinable, and analytically calculable from field solutions in a manner which resembles that used for the simple inductor example above. However, since the transformer is a multi-winding device, both the questions that need to be asked and their answers tend to be a little more complicated.

The Conventional Equivalent Circuit

Many small transformers are constructed by placing a pair of windings around the center leg of a magnetic core, as indicated in Fig. 9(a). The windings are placed over each other, usually with the primary winding

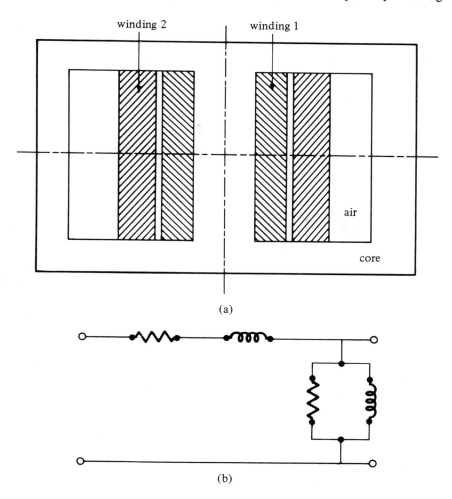

(a)

(b)

Figure 9. (a) A small two-winding transformer, with (b) its conventional equivalent circuit representation.

nearer the core center. The core is a built-up stack of E- and I-shaped stamped steel laminations, laid with the E and I pieces alternating in orientation, so that a closed but gapless magnetic structure is achieved. This structure resembles the simple inductor of Fig. 1, except for two details: the transformer comprises two windings while the inductor has one, and the inductor frequently (though not always) incorporates one or more air gaps in its magnetic path.

The transformer is structurally similar to a simple inductor, and the two devices are analyzed by techniques similar in principle. The assumption is once again made that very little leakage flux escapes the core, practically all the flux being confined to either the core itself and to the window space which contains the windings. End leakage will be ignored in a first analysis; some compensation for this approximation is possible by giving the air in the window space a relative permeability a little higher than unity. In a two-dimensional representation of the transformer, there are two symmetry planes, as indicated in Fig. 9(a); only one-quarter of the transformer need therefore be given an explicit representation for analysis.

Like many other electromagnetic devices, transformers are very frequently used as component parts in large, complex systems. They are commonly described by equivalent circuits, so that circuit descriptions of whole systems can be composed out of the equivalent circuits of their various component devices. Several equivalent circuits are available for transformers, all of roughly similar complexity. The most common equivalent circuit representation comprises two parts: a perfect (lossless and leakage-free) transformer with the correct turns ratio and the T-circuit shown in Fig. 9(b). The latter contains the circuit parameters which set the real device apart from an ideal transformer.

One reason why the T-shaped transformer equivalent circuit of Fig. 9(b) is in widespread use is that its components correspond fairly closely to the various physical phenomena in the transformer. The reactance in the vertical leg is essentially similar to the single reactance in Fig. 2: it accounts for the energy stored in the iron core. However, in the inductor the stored energy associated with leakage flux could be lumped in with the energy stored in the core; in the transformer it cannot, because there do exist measurements affected by the one and not the other. The series resistances in the horizontal legs of the T are the same wire resistances as in the inductor equivalent circuit of Fig. 2 above; the difference is that there are two windings and hence two resistances, while the inductor has only one. The shunt resistance in the vertical leg is a reasonably good representation of the power loss in the iron core. Identifying equivalent circuit elements with physical phenomena in this fashion is not strictly accurate; but the error involved is minor in most cases, because the numerical values of the vertical and horizontal branch circuit components in Fig. 9(b) differ by at least one, and occasionally as much as three, orders of magnitude.

The ideal transformer shown in Fig. 9(b) is usually omitted in analysis of systems containing the transformer. It is usual among transformer engineers to refer all quantities to the source-side winding, and to speak of the equivalent circuit as if it did not contain an ideal transformer or, what is the same thing, as if the turns ratio of the transformer were exactly unity. The voltages and currents that appear in the equivalent networks must then be scaled up and down, respectively, by the turns ratio to bring them into accord with the actual values. The minor inconvenience of voltage, current, and impedance referral, however, is richly compensated by the simplification that results when writing and solving circuit equations. This convention will be adhered to in what follows.

Short-Circuit Parameters

In the laboratory, the leakage reactance and winding resistance of the transformer are measured by means of the so-called *short-circuit* test. In this test, the secondary (load) terminals of the transformer are short-circuited, and the primary is fed with a low-voltage source so as to draw the rated current. Viewed in terms of the equivalent circuit of Fig. 9(b), the test arrangement amounts to short-circuiting the right-hand terminal pair. Very little current then flows in the shunt (magnetizing) branch of the circuit, since the impedance of this circuit branch is normally very high; typically, the shunt branch current might amount to 1–5% of the total. Primary-side measurements of voltage, current, and power therefore relate to the leakage reactance and winding resistance only; indeed, they provide a reasonable way of measuring these quantities.

Simulation of the classical short-circuit test is possible using a CAD system, but a better idea is to substitute another test, which might be called the *bucking test*; it is more accurate but much more difficult to perform in the laboratory. In the classical short-circuit test, a sinusoidal current source is connected to the left-hand terminals of Fig. 9(b) and the right-hand terminals are short-circuited. In the simulated test, two current sources are employed, one connected to the left and one to the right terminal pair. The two sources are identical, so that the current flowing into the left terminals is exactly equal to the current flowing out at the right. The magnetizing (vertical) branch of the equivalent circuit must then carry exactly zero current in the bucking test, in contrast to the conventional short-circuit test in which its current is approximately zero. For determining the horizontal branch parameters of Fig. 9(b), the bucking test is clearly superior because it eliminates the magnetizing branch completely. It is easy to implement in a CAD system, where it is only necessary to prescribe equal currents; in the laboratory, on the other hand, it is perfectly feasible in principle but very hard to carry out in practice. In the transformer core and window, such a test produces the flux distribution shown in Fig. 10. It will be seen that all the flux lines link at least part of one winding or the other. There are no flux lines which close in the iron

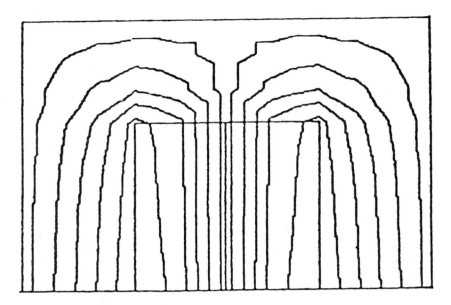

Figure 10. Flux plot for transformer on short circuit.

alone, showing that the two winding currents are exactly equal, with no net magnetomotive force applied to the core. To put the matter another way, there is zero net flux linkage of the secondary and primary windings.

The leakage inductances of the transformer can be determined by calculating the total stored energy associated with the solution of Fig. 9(b), using the technique of equation (27). Alternatively, the flux linkage technique of equation (21) may be applied; there should be no difference in the results, since almost all the stored energy will reside in the air. The flux density in the iron core is invariably very low at short circuit in a well designed transformer, and the energy stored in the iron core is therefore tiny compared to the energy stored in the leakage field in and around the windings. In fact, the short-circuit test simulation is very frequently performed on the assumption of infinite iron permeability, so that the energy stored in the core is not taken into account at all.

When short-circuit tests are performed in the laboratory, only the combined reactance value of the two windings (referred to the primary) is determined. It is conventional to apportion half the leakage inductance to the primary, half to the secondary winding. This apportionment may seem arbitrary but is often unavoidable. The single measurement made in the short-circuit test can only determine a single reactance, the combined total of primary and secondary leakages; separating them requires at least one further laboratory experiment. But the CAD system user, unlike the laboratory experimenter, obtains a full field solution from the (simulated) experiment, not merely one or two terminal values. He is therefore able to carry out additional simulated measurements by performing further mathematical manipulations on the fields; furthermore, he is free to devise simulated measuring instruments which would be very difficult to realize in the laboratory. This flexibility permits separation of primary from secondary leakage inductances in cases where the windings are not symmetrically disposed and where a simple half-and-half split may not be appropriate. It is only necessary to write the total stored energy in the form

$$W = \frac{1}{2} \int_{S_p} A \, J \, dS + \frac{1}{2} \int_{S_s} A \, J \, dS, \qquad (31)$$

where S_p is the cross-sectional area of the primary winding, S_s that of the secondary winding. In accordance with equation (20), this energy may be written in terms of the primary and secondary flux linkages $(n\phi)_p$ and $(n\phi)_s$ as

$$W = \frac{1}{2} \left[(n\phi)_p + (n\phi)_s \right] i. \qquad (32)$$

Since no mutual flux linkages are shared between primary and secondary in the bucking test, it is proper to rewrite the energy W in terms of the primary and secondary leakage inductances L_p and L_s as

$$W = \frac{1}{2} \left[L_p i + L_s i \right] \qquad (33)$$

showing that the individual leakage inductances are separately computable. The primary leakage inductance is thus, by (22),

$$L_p = \frac{1}{i^2} \int_{S_p} A \, J \, dS \qquad (34)$$

and a similar expression holds for the secondary. The leakage inductances are separable, in other words, by measuring the primary and secondary flux linkages separately, a task not particularly difficult in the simulated bucking test but impossible in laboratory practice.

It should be noted that the winding resistances in the equivalent circuit cannot be calculated directly from the magnetic field distribution. As with the simple inductor, a value can be deduced from the length of wire used and its conductivity.

Magnetizing Inductance

The magnetizing reactance and core loss resistance of transformers are determined in the laboratory by performing an *open-circuit* test. In this test the secondary terminals are left open-circuited and the primary is excited at rated voltage. Measurements of the primary voltage, current, and power, together with the results from the short-circuit test, allow the calculation of the magnetizing reactance and the core loss resistance. Since the magnetizing impedance of a well designed transformer is very considerably greater than the shunt impedances, the results of the open-circuit test are quite often used to calculate the magnetizing impedance directly, ignoring any correction for the leakage and winding resistance.

The open-circuit test is not only easy to perform in the laboratory, because the transformer is run at rated voltage without load, but it is also easy to model analytically. Leaving the secondary winding open-circuited simply requires that it is ignored in setting up the problem! In fact this test is identical to the simulation, already treated, of a simple inductor, since only one winding is taken into account in the modelling. With the primary winding on the left, the transformer of Fig. 9(a) exhibits the open-circuit flux distribution shown in Fig. 11.

Since the open-circuit test is performed at rated voltage, not at rated current, its simulation involves a few difficulties. First, most present-day CAD systems work best with prescribed currents rather than voltages. Secondly, the *magnetizing reactance* is ill defined, since most transformers are operated at sufficiently high saturation levels to cause the magnetizing current to contain a significant proportion of harmonics. With sinusoidal applied voltage, but nonsinusoidal current, the inductance—and hence the reactance—of the primary winding can be defined in several ways. This problem does not arise in the leakage calculations associated with the short-circuit test, since much of the magnetically stored energy is stored in air. The contrary is true for the open-circuit test: almost all the magnetic energy is stored in the iron core, which is after all nonlinear.

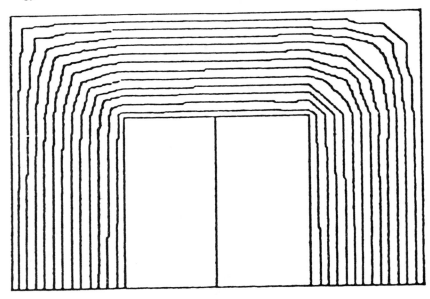

Figure 11. Flux plot for transformer on open circuit.

Like many other simulations, the open-circuit test can be carried out in a careful and fairly precise way, or it can be done quickly but less accurately. The slow and precise way proceeds as follows. The range of instantaneous magnetizing currents, up to the largest peak current likely to be encountered, is determined by any reasonable means (which may include an educated guess). A set C of N current values, which may number about ten, is chosen so as to span the range,

$$C = \{i_k, \ k = 1, \ldots, N\}. \tag{35}$$

The open-circuit field problem is solved for each one of these values, thereby producing a set A of vector potential distributions $A_k(x,y)$,

$$A = \{A_k(x,y), \ k = 1, \ldots, N\}. \tag{36}$$

The primary flux linkages are next computed for each solution, using equation (20),

$$(n\phi)_k = \frac{1}{i_k} \int_{S_p} A \, J \, dS. \tag{37}$$

Since every flux linkage value $(n\phi)_k$ in this set corresponds to a specific value i_k, the two sets taken together describe a function which assigns a specific flux linkage value to each current value, and vice versa:

$$(n\phi)_k = n\phi(i_k). \tag{38}$$

The resulting curve of flux linkages against current will roughly resemble a B-H curve. However, it will not be proportional to the B-H curve of the core material, because the flux distribution in the core actually changes as the material saturates.

When an open-circuit test is carried out in the laboratory, the primary winding is excited with a sinusoidal voltage, say

$$e(t) = E \cos \omega t. \tag{39}$$

By Faraday's law, the primary flux linkages must therefore vary in a sinusoidal fashion,

$$n\phi = -(1/\omega) E \sin \omega t, \tag{40}$$

so that the sequence of time instants t_k may be determined at which the tabulated values of flux linkage are reached during a quarter cycle:

$$t_k = -(1/\omega) \arcsin \left[-\omega n\phi(i_k)/E \right]. \tag{41}$$

But this relationship creates a set T of time values,

$$T = \{t_k, \ k = 1, \ldots, N\}, \tag{42}$$

which correspond to the current values C of equation (35). If these current values are plotted against time, the magnetizing current waveform is obtained. It may be written as a Fourier series, as detailed further below. Because the total current waveform has quarter-cycle symmetry and begins at $i(0) = 0$, the series contains odd sine terms only:

$$i(t) = \sum_{m=1}^{\infty} I_m \sin (2m-1)\omega t.$$ (43)

The magnetizing inductance is then calculated by taking the fundamental components only,

$$L_{mag} = \frac{E}{\omega I_1}.$$ (44)

The fundamental-frequency component definition of inductance is probably the best to choose in this application, because a main use of small transformer equivalent circuits is in the calculation of fundamental-frequency currents in systems containing transformers. Getting the fundamental components right is therefore likely to be important, more so than, say, finding the stored energies with high accuracy. The same cannot be said, however, of applications involving transients. Most transient phenomena involve transfer of flux linkages or stored energies and therefore should use inductances based on total quantities, rather than specific Fourier series components.

A quick but less accurate estimate of the magnetizing reactance can be obtained by means of a single static solution. The primary winding current is set at the root-mean-square value expected, and its flux linkages are computed from the field solution. A value of terminal voltage is then obtained, on the assumption that the value obtained for flux linkages is the root-mean-square value also. If the resulting voltage value is within a few percent of the rated voltage, the magnetizing reactance value is probably accurate to a few percent also.

It may be worth noting that no attempt has been made in either of the above methods to allow for the primary leakage reactance. Since both calculations actually deal with *terminal* voltage, the reactance values obtained represent not the magnetizing reactance, but rather the sum of magnetizing reactance and leakage reactance. To obtain a value for the magnetizing reactance alone, it suffices to subtract the leakage reactance value obtained from the bucking test simulation described above.

Harmonic Analysis

The harmonic content of magnetizing current is a quantity of considerable interest to transformer designers as well as to the system engineer using transformers as circuit components. If hysteresis is absent, as has been assumed in the above, then the magnetizing current will have quarter-cycle symmetry; that is,

$$i(\omega t + \pi) = - i(\omega t). \tag{45}$$

The Fourier series of $i(\omega t)$ can then contain no odd terms; and

$$i(\omega t + \pi/2) = - i(-\omega t) \tag{46}$$

so there can be no cosine terms. The Fourier series of $i(\omega t)$ therefore has the form already alluded to,

$$i(t) = \sum_{m=1}^{\infty} I_m \sin (2m-1)\omega t. \tag{43}$$

In any real situation, the true current $i(t)$ cannot be known, since an infinite amount of data would be needed to determine all Fourier series components. Instead of (43), the practical analyst must remain content with the approximate current $i_M(t)$, which contains only M terms of the series,

$$i_M(t) = \sum_{m=1}^{M} I_m \sin (2m-1)\omega t. \tag{47}$$

The number M is of course limited to at most N, the number of data points contained in the sets of current and time values, equations (35) and (42). This restriction is not particularly severe, for there is rarely much interest in harmonics beyond the fifth or (in some three-phase systems) eleventh. That is to say, M is limited to 3, or 6 at most, while present CAD systems make it easy to compute solutions for, say, $N = 10$.

Since the number N of data points available is frequently larger than the number M of Fourier series terms required, it is best to compute the limited number of terms by a least-squares approximation. To do so, the series (47) is formally evaluated at each of the N time instants t_k,

$$i_M(t_k) = \sum_{m=1}^{M} I_m \sin (2m - 1)\omega t_k. \tag{48}$$

The coefficients I_m are chosen so as to make the squared difference between left- and right-hand sides of (48), taken at all the N time instants, as small as possible. That is to say, the I_m are chosen so as to make first derivatives vanish:

$$\frac{\partial}{\partial I_n} \sum_{k=1}^{N} [i(t_k) - i_M(t_k)]^2 = 0. \tag{49}$$

Substituting (48) into (49) and differentiating, a set of simultaneous equations is obtained. For every value of m,

$$\sum_{n=1}^{M} \sum_{k=1}^{N} \sin(2n-1)\omega t_k \ \sin(2m-1)\omega t_k \ I_n$$

$$= \sum_{k=1}^{N} \sin(2m-1)\omega t_k \ i(t_k) \tag{50}$$

must hold. These equations are easily recast in matrix form for numerical solution, noting that the inner summation (with index k) on the left-hand side collapses the trigonometric functions into a symmetric square matrix with indices n,m.

The above discussion assumes that hysteresis is absent. Indeed, no other assumption can at present be fruitfully made, for no general-purpose CAD systems now available can treat hysteretic behavior in solving field problems. The analysis given here is most accurate at high saturation levels, when the hysteresis loop width is small compared to the excursion in magnetic field.

Loss Estimation

The core loss resistance to be included in the transformer equivalent circuit may be estimated roughly from a single static field solution. The estimating technique is simple: the local power loss density p (usually given on a per unit mass basis) is weighted by the mass density $m(x,y)$ of the stacked laminations and integrated over the volume of magnetic material so as to obtain the total loss P. Thus

$$P = Z \int_{S_{iron}} p \ m(x,y) \ dS, \tag{51}$$

where Z represents the stacking depth of the core iron, and S_{iron} is the area occupied by iron in the transverse (x-y) plane. The loss density $p(x,y)$ of course depends on the magnetic events at the point (x,y). It is usually expressed as a function of peak flux density,

$$p = p(B_p). \tag{52}$$

The function $p(B_p)$ must be obtained either from the iron supplier or by direct experimental measurement. Most suppliers of magnetic materials provide core loss curves for sheet and strip stock, in watts per kilogram or pound as a function of the peak flux density. By referring to mass rather than volume, any required stacking factors are automatically taken care of by the mass density $m(x,y)$ in equation (51). Curves are usually available for sinusoidal flux density variation, at 50 Hz and 60 Hz for the thicker materials and 400 Hz for the thinner grades. Data for other frequencies or working conditions are furnished only infrequently.

Since the loss curves refer to peak flux density, the loss is estimated from a field solution obtained for the open-circuit test with the primary winding carrying peak current. This solution is identical to one obtained for determining the magnetizing reactance. The potential values in it are differentiated, to yield the corresponding flux density B_k,

$$\mathbf{B}_k = \text{curl}\,(\mathbf{1}_z A_k).\tag{53}$$

These represent peak values, since they are derived from the peak values of potential. The loss density curve supplied by the manufacturer is then used to determine the loss density everywhere in the core, as in (52), and the total loss is obtained by integration over the whole core, as in equation (51).

The core loss resistance R to be included in the transformer equivalent circuit may be calculated from the estimated core loss P. By simple circuit theory, the value of R is

$$R = \frac{E_{rms}^2}{P}.\tag{54}$$

If voltage drop in the primary resistance and leakage inductance is neglected—a reasonable approximation for the open-circuit test—then E_{rms} is the root-mean-square voltage at the primary terminals. Because the transformer may be assumed to be excited by a sinusoidal voltage, (54) may be rewritten in terms of the peak voltage E_p,

$$R = \frac{E_p^2}{2P}.\tag{55}$$

But the peak terminal voltage is related to the primary flux linkages $n\phi$ by equation (40). Hence the value for core loss resistance is finally obtained as

$$R = \frac{\omega(n\phi)_p^2}{2P}.\tag{56}$$

It should be noted that the loss as estimated using the standard core loss curves refers only to the loss incurred by alternating flux, not rotational flux density. Rotating fluxes arise in many devices through the redistribution of flux which arises from saturation. The distribution of fluxes in a transformer core, or indeed in any magnetic device, is clearly not the same at high and low saturation levels. As the exciting current is raised, regions of high flux density saturate first, crowding flux into areas of lower density. At high saturation levels, flux densities therefore tend to be much more uniform than at low total flux values, and the local direction of flux is not the same in the two cases. As a transformer on open-circuit test traverses the alternating current cycle, the flux density vector $\mathbf{B}$ at most points of the core iron not only varies in magnitude but also changes in direction as flux redistribution takes place. Hence the vector $\mathbf{B}$ rotates in direction as its magnitude alternates; its tip generally traces an elliptic path. Iron loss curves, however, make no allowance for rotating components in the flux density. Estimates of loss are therefore usually less accurate than estimates of the corresponding magnetizing reactance.

Mutual Inductances

Many phenomena in electromagnetic devices are conveniently described in terms of mutual inductance values, and many designer-years are spent calculating values of mutual inductances. Ways of extracting mutual inductances from field solutions therefore merit more than passing mention.

Mutual Inductances and Energy

Since there are at least two reasonable bases on which to define self-inductance—stored energy and flux linkages—it should not be surprising that mutual inductances can be similarly defined.

One note of caution is perhaps in order: the computation of inductances, like the computation of almost anything else in a CAD system, should be undertaken only as a *simulation* of experimental measurements. A corollary is that no attempt should ever be made to calculate quantities which cannot be defined in terms of a physically feasible experiment. While this principle may seem obvious, it is very frequently violated by classical engineering electromagnetics. For example, the concept of *internal inductance* of a conductor is enshrined in textbooks and has been so for a hundred years. It is well known and calculable for a round wire, for which analytic solutions of the skin-effect problem exist. It is not calculable for other, more complicated shapes—not because of shortcomings in the available processes of calculation, but because the classical definition of internal inductance itself makes no sense for other conductor shapes. (It assumes that one flux line coincides with the conductor surface, partitioning all other flux lines into those *internal* to the conductor and those *external* to it. Such a separatrix flux line can only exist if the conductor is round.) Many of the classically useful concepts and quantities of electric machine engineering in particular are imprecise because their definitions contain built-in presuppositions about geometric shapes: examples are end-turn inductance, slot leakage inductance, and zig-zag leakage. If there is any doubt at all about some well established parameter, the one sure test is to invent a physical experiment for measuring the relevant quantity. In effect, the invention of such an experiment redefines the quantity in experimental terms valid for all cases. Even if the measurements are impractical to carry out, their simulation may be quite straightforward.

Defining mutual inductance in terms of stored energy is a secure procedure, for the stored energy is a quantity easily measured, at least in principle, by an integrating wattmeter. The mutual inductance in a two-winding system may be determined by a difference measurement involving two experiments, one with the two winding currents oriented to have their fluxes adding, the other with the fluxes bucking each other, as in Fig. 12. It is of course not necessary to know at the outset which is which; it is only essential to conduct two experiments, with connections to one winding reversed relative to their orientation in the other experiment, i.e., to have the winding currents i_1 and i_2 in one case, i_1 and $-i_2$ in the other. Altering the order of experiments will make the mutual inductance have either a positive or a negative sign.

Let W_1 and W_2 be the stored energies in the two experimental cases. By simple circuit theory, they are related to the self-inductances L_{11}, L_{22} and the mutual inductances M_{12}, M_{21} through the relationships

$$W_1 = \frac{1}{2} \left(L_{11}\, i_1^2 + M_{12}\, i_1\, i_2 + M_{21}\, i_2\, i_1 + L_{22}\, i_2^2 \right) \tag{57}$$

and

$$W_2 = \frac{1}{2} \left(L_{11}\, i_1^2 - M_{12}\, i_1\, i_2 - M_{21}\, i_2\, i_1 + L_{22}\, i_2^2 \right). \tag{58}$$

By subtracting, the difference in stored energies is obtained. There immediately results

$$M_{12} + M_{21} = \frac{W_1 - W_2}{i_1\, i_2}, \tag{59}$$

and since mutual inductances are reciprocal, $M_{12} = M_{21} = M$,

$$M = \frac{W_1 - W_2}{2\, i_1\, i_2}. \tag{60}$$

The procedure as given above requires measurement or computation of stored energy, followed by subtraction to find the energy difference. If the energies W_1 and W_2 are not very different, the result can be considerably in error. This *subtraction of elephants* difficulty arises no matter whether the data are obtained by direct physical experimentation or by computer simulation: if the difference looks like small mice, the result is of doubtful accuracy. To be more precise: as many significant figures will be lost in the subtraction as there are similar leading digits in W_1 and W_2. It is always well to inspect results, including intermediate results, from time to time as calculations proceed and to bear this potential difficulty in mind.

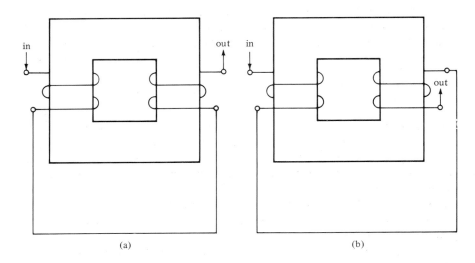

Figure 12. Two windings connected together for purposes of mutual inductance determination: (a) series aiding, (b) bucking.

The stored energy calculations required in the above simulation may be carried out by exactly the same techniques as followed in the simple inductor case. The appropriate equations to use are generally (25) and (26), but circumstances may allow simplifications arising from magnetic linearity or geometric shape.

Mutual Inductance and Flux Linkage

Like self-inductances, mutual inductances may be defined in terms of flux linkages. Such a definition leads to quite direct and simple calculations, possibly easier than the energy-based approach. In linear problems, the two definitions again coincide, so the results obtained are identical to within the error inherent in numerical computation; the choice of method is therefore a matter of convenience. In nonlinear cases, however, flux linkages and stored energy must lead to different results because they reflect two different definitions of inductance.

In terms of flux linkages, the mutual inductance M_{12} is defined as *the flux linkages of winding 1 caused by the current in winding 2*, divided by the current causing them:

$$M_{12} = \frac{(n\phi)_{12}}{i^2},\tag{61}$$

As already discussed in connection with equation (20), flux linkages are best determined with reference to currents, so that any nonuniformity in the winding density is automatically introduced as a weighting factor. To determine the flux linkages in winding 1, it is therefore best to imagine that a very small current i_1 is made to flow in that winding, one small enough to cause negligible change in vector potential. This small current will nevertheless cause a current density J_1 to exist in winding 1. Generalizing on equation (20) slightly, the required flux linkage may then be calculated as

$$(n\phi)_{12} = \int A \left[\frac{J_1}{i^2} \right] dS. \tag{62}$$

Combining (61) and (62), the mutual inductance is thus

$$M_{12} = \frac{1}{i_2} \int A \left[\frac{J_1}{i_1} \right] dS. \tag{63}$$

The region of integration here may be the cross-sectional area of winding 1, or any larger area. A larger area is perfectly acceptable, since the current density J_1 will vanish in any portion not actually occupied by winding 1. No extra contribution will accrue to the integral even if the area is too large. As a practical matter, the area is frequently chosen to include the entire problem region, thereby reducing work in defining the region of integration.

When the calculation of (63) is actually carried out in a CAD system, there is no need to keep i_1 small, provided the vector potential A as used in the calculation is the vector potential that results from solving the field problem with winding 2 excited, but without current in winding 1. In other words, there is no need to risk numerical error by choosing i_1 to be very tiny; any convenient value will do.

Force Calculations

Many of the uses to which magnetic devices are put may be classified as electromechanical; that is, they are used to convert energy between electrical and mechanical forms. Indeed, mechanical force production is the major reason for the existence of devices such as actuators and electric motors. Consequently, the end product of a magnetic field analysis may well be the evaluation of the mechanical force produced by the device and its variation with changes in excitation or position. The principles involved are introduced and discussed here with reference to illustrative examples.

A Magnetic Bearing

Permanent magnet bearings are employed in some watt-hour meters. Such meters are in essence small electric motors designed to measure electricity consumption, which must have very consistent performance in order to meet legal and commercial needs. The usual type of meter includes a rotating aluminum disk, driven by eddy currents caused by two excitation coils—one coil intended to measure instantaneous current, the other instantaneous voltage. Integration over time is performed mechanically, by having the disk actuate a rotating counter. For satisfactory operation the rotating disk mechanism should exhibit little friction, and the bearing characteristics should show negligible change with diurnal and seasonal weather variations as well as over the lifetime of the device, which is likely to be measured in decades.

To meet stability and consistency requirements, one type of watt-hour meter incorporates a permanent magnet suspension system with an auto-compensating magnetic shunt to allow for the temperature variation of permanent magnets. Such a configuration is represented in Fig. 13. A compensating shunt (flux diverter) bypasses part of the magnetic flux of the permanent magnet. With an appropriate choice of materials and dimensions, the amount of flux diverted can be made to vary with temperature in the sense opposite to the permanent magnet material. For example, an increase in total flux, resulting from temperature variations, is compensated for by an increase in the fraction of flux diverted. In this way, the suspension height can be kept practically constant, and the magnetic bearing can produce a watt-hour meter with precision that rivals mechanical suspension, but without the wear that results from mechanical contact.

For the magnetic bearing of Fig. 13, the field at nominal design temperature is shown in Fig. 14. The analysis shown was carried out using explicit current-carrying coils rather than intrinsic permanent magnet models, and the coil sides (which actually are empty air space) are clearly visible in Fig. 14. The flux diverter accounts for very little leakage at the design temperature, which corresponds to summertime operation.

Although it is necessary to obtain lifting force values as a final analytic result, preliminary examination of possible designs may require no more than simple visual inspection of field distributions. Often an extremely

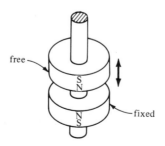

Figure 13. A simple vertical-shaft magnetic bearing used in a watt-hour meter, based on permanent magnets.

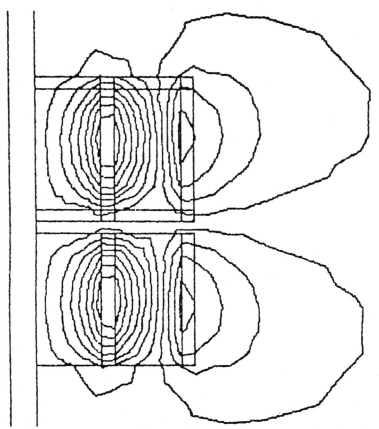

Figure 14. Magnetic field (flux lines) of the permanent magnet bearing shown in Fig. 13.

useful indication of the behavior of a magnetic system can be obtained in this qualitative way. In Fig. 14 the flux lines are crowded together in the air gap between the two magnets and are directed horizontally. This strong tangential component of field in the air gap suggests that there is a substantial repulsion force between the two parts of the bearing. The magnitude of the force can be gauged very roughly by observing both the density of the lines and how horizontal they appear to be.

The actual value of the force in a magnetic system may be determined in several ways. Since there is in reality only one force, all the ways should produce the same result, provided there is no computational error. However, because of the numerical approximations which have been made in the solution of the field equations and the distribution of the force itself, the several methods may not give identical answers. In fact, the most appropriate method to use may well depend on the device being analyzed. Some of the more common approaches and their pitfalls are outlined below.

The Method of Virtual Work

The force exerted on a body may be evaluated by determining the work done when it is slightly displaced from its rest position. In the absence of frictional losses, this work must equal the change in the energy stored in the entire electromechanical system or device. Thus, if the stored energy is evaluated for the device in two positions separated by a small displacement, then the difference in energy divided by the distance will give a value for the force. Suppose, for example, that the stored energy is W_1 in position 1 and W_2 in position 2, the two positions being separated by a displacement x_{12}. The force required is then given by

$$F = \frac{W_2 - W_1}{x_{12}}. \tag{64}$$

This approach to force calculation is widely employed in mechanics, where it is known as the *method of virtual work*. It evidently relies on the assumption that the force does not change significantly during the motion and thus is valid for small displacements only. In the completely general case where displacements may take place in various directions, the force is a vector quantity, given by

$$\mathbf{F} = \text{grad } W(x), \tag{65}$$

where $W(x)$ is the stored energy, viewed as a position function of the vector displacement x. Equation (64) may be regarded as a special case of (65), for one-dimensional movements.

The lifting force for the magnetic bearing described here was computed by the method of virtual work. Since lift force is the desired quantity, the upper magnet of the bearing was displaced so as to halve the vertical air gap between magnets, from 2.0 mm to 1.0 mm. (By way of comparison, the bearing magnet diameter is 18 mm.) The force is likely to change more or less linearly with distance, in view of the large bearing magnet diameter, so that the value obtained is likely to represent a good estimate for the actual force at an air gap of 1.5 mm. Values for the stored energies in the two positions and the force are as follows:

Stored energy with 2 mm gap	=	2.9476 mJ
Stored energy with 1 mm gap	=	2.8892 mJ
Change in stored energy	=	0.0584 mJ

and the force follows directly as

$$\text{Force} = \frac{0.0584 \text{ mJ}}{1.000 \text{ mm}} = 0.0584 \text{ N}.$$

It may be noted that the energy difference in this case is about 2% of the stored energy itself. In other words, nearly two leading significant figures are alike in the two energies, and nearly two significant figures will therefore be lost to roundoff error in the subtraction. If the energies themselves can be relied on to about five figures, then three-figure accuracy of the force is all that can be hoped for.

The figures above are fairly typical for a calculation based on the virtual work method, although in this case a comparatively large displacement can be made. In systems where only a very small displacement is possible and the forces are likely to be small, the change in stored energy may be a fraction of a percent of the total energy. The difficulty of *subtraction of elephants* then arises once again, just as it did with equations (59)–(60). There are two ways it can be alleviated: wisdom in solution of the field problems and wisdom in differentiation.

When forces (or mutual inductances!) are computed by subtracting energies, accuracy can be considerably enhanced if the error inherent in the two energies themselves is similar. If, say, W_1 as computed contains some error e_1, as $W_1 + e_1$, and W_2 is similarly computed as $W_2 + e_2$, then obviously

$$(W_2 + e_2) - (W_1 + e_1) = (W_2 - W_1) + (e_2 - e_1). \tag{66}$$

The error in the energy difference will certainly be no larger than the error in either energy; indeed it will generally be smaller, if it can be guaranteed that both e_2 and e_1 have the same sign. Such a guarantee is easily furnished if the field problems underlying W_1 and W_2 are both solved using energy-minimizing finite element methods. Fortunately, such methods are by far the most popular ones in present-day CAD systems. The best results will clearly be obtained if e_2 and e_1 are not only of the same sign but of similar magnitude. Their magnitudes are very often mainly affected by the discretization error that arises from the finite element mesh. If W_1 and W_2 are solutions computed on the same finite element mesh, or very similar meshes, errors are likely to be similar and the energy difference therefore much more accurate than the simple rules of thumb would indicate.

The results obtainable by virtual work calculations can sometimes be improved by employing not the simple equation (64), but its more general version (65). Instead of computing energy at the two ends of the displacement x_{12}, a set of several stored energies can be found at several points along the displacement vector. If a smooth curve is fitted to these several values, and differentiation performed according to (65), considerable error smoothing can be achieved at the price of increased computation.

Maxwell Stresses

The second most common approach to determining electromechanical forces is that known as the *Maxwell stress tensor* method. In contrast to the virtual work technique, which employs a volume integral to determine the stored energy, the Maxwell stress approach computes local stress at all points of a bounding surface, then sums the local stresses by means of a surface integral to find the net force.

The Maxwell stress tensor method may be derived from the elementary force density expression which relates the force density vector $\mathbf{f}$ to the flux density $\mathbf{B}$ and the current density $\mathbf{J}$ by

$$\mathbf{f} = \mathbf{J} \times \mathbf{B}. \tag{67}$$

This expression is derived in theoretical electromagnetics from the fundamental force relationship between two moving charges, and represents the magnetic portion of the *Lorentz force*.

The expression given in equation (67) describes a force density vector, which possesses components in each of the coordinate directions and has dimensions of force per unit volume. Thus the force on a body in a particular direction may be found by integrating the appropriate component of the force vector over the entire volume. It is common, however, to reduce the volume integral described above to a surface integral in order to simplify the overall force calculation. If a substitution is made for $\mathbf{J}$ in equation (67) by using Maxwell's equations, the force expression becomes

$$\mathbf{f} = \nu \, \mathbf{B} \times \operatorname{curl} \mathbf{B}. \tag{68}$$

The left-hand side of this equation may be expanded into three components. The x directed component is typical and has the following form:

$$f_x = \nu \, B_z \, \frac{\partial B_x}{\partial z} - \nu B_z \, \frac{\partial B_z}{\partial x} - \nu B_y \, \frac{\partial B_y}{\partial x} + \nu B_y \, \frac{\partial B_x}{\partial y}. \tag{69}$$

If a term $\nu B_x(\partial B_x/\partial x)$ is simultaneously added to and subtracted from equation (69), and the identity

$$\frac{\partial}{\partial x} (B_x)^2 = 2B_x \, \frac{\partial B_x}{\partial x} \tag{70}$$

is used, then the force component becomes

$$f_x = \nu \left[\frac{1}{2} \frac{\partial}{\partial x} (B_x)^2 + B_z \, \frac{\partial B_x}{\partial z} + B_y \, \frac{\partial B_x}{\partial y} \right.$$
$$\left. - \frac{1}{2} \frac{\partial}{\partial x} (B_x^2 + B_y^2 + B_z^2) \right]. \tag{71}$$

Some further manipulation gives

$$f_x = \left[\frac{\partial}{\partial x} \left(B_x^2 - \frac{1}{2} \mid B \mid^2 \right) + \frac{\partial}{\partial y} (B_x B_y) \right.$$
$$\left. + \frac{\partial}{\partial z} (B_x B_z) - B_x \operatorname{div} \mathbf{B} \right]. \tag{72}$$

Since div $\mathbf{B} = 0$, the last term disappears and the remaining expression may be recognized as the divergence of a vector $\mathbf{f_x}$, whose components are

$$f_{xx} = \nu[B_x^2 - \frac{1}{2} \mid B \mid^2], \tag{73}$$

$$f_{xy} = \nu B_x B_y,$$

$$f_{xz} = \nu B_x B_z. \tag{73}$$

A similar development holds for each of the other force components ($\mathbf{f_y}$ and $\mathbf{f_z}$). Thus the force expression, equation (67), may be written as the divergence of some *tensor* $\mathbf{T}$:

$$\mathbf{f} = \int_U \mathbf{J} \times \mathbf{B} \, dU = \frac{1}{\mu} \int_U \text{div} \, \mathbf{T} \, dU. \tag{74}$$

Making use of the divergence theorem, this volume integral may be reduced to a surface integral. Then the force becomes

$$\mathbf{f} = \frac{1}{\mu} \oint_S \mathbf{T} \cdot \mathbf{dS}, \tag{75}$$

where the surface vector $\mathbf{dS}$ is taken as the outward normal on S.

The tensor $\mathbf{T}$ defined above has the dimensions of stress and is commonly known as the *second Maxwell stress tensor* or the *magnetic stress tensor*. A more complete derivation of this tensor may be obtained by considering both the magnetic and electric components of the Lorentz force. In this case, an electric stress tensor may be defined in addition to the magnetic one described here. The electric stress tensor has essentially the same form as its magnetic counterpart with $\mathbf{B}$ replaced by $\mathbf{E}$ and reluctivity replaced by permittivity. This form is useful for force calculations in electrostatic fields. Furthermore, when both electric and magnetic forces are considered together a link between the two fields appears which may be recognized as the *Poynting vector* representing power flow within the volume.

The complete magnetic stress tensor $\mathbf{T}$, written out in full, has the following form:

$$\mathbf{T} = \begin{vmatrix} (B_x^2 - \frac{1}{2} \mid B \mid^2) & B_x B_y & B_x B_z \\ B_y B_x & (B_y^2 - \frac{1}{2} \mid B \mid^2) & B_y B_z \\ B_z B_x & B_z B_y & (B_z^2 - \frac{1}{2} \mid B \mid^2) \end{vmatrix}. \tag{76}$$

Equation (76) provides the local values of all components of magnetic stress along each of the coordinate axes. The rigid body force acting on an object is obtained by integrating these components over its bounding surface, as in equation (75).

It should be remembered that each stress component is in fact a vector and thus the dot product with the normal should be applied to each term of the appropriate horizontal row in turn. In two dimensions the 3×3 matrix reduces to 2×2, i.e. the top left hand corner of the tensor, and the surface integral becomes a contour integral. Thus, if the surface is parallel to the x-coordinate axis in a two- dimensional system, the x-directed component of force is given by $B_x B_y$ and the y directed force is $(B_x^2 - |B|^2/2)$. In the case of two blocks of iron facing each other, such as might be encountered in an electromagnet where all the flux is directed between the poles, there is no sideways force because B_x is zero. The attractive force between the two poles is then given by $B_y^2/(2\mu_0)$, which may be recognized as the conventional expression for the force between two magnetic poles.

To determine the forces on a rigid body, the surface of integration S should of course encompass the body. In principle, S should be the surface of the body itself. In computational practice, it is often found convenient to place this surface in the air region surrounding the machine part or other object on which the force is to be found. In effect, a "piece of air" is therewith attached to the rigid body, as if it were a solid. Since the air carries no currents and has no magnetic properties different from free space, there is no harm in this hypothetical attachment.

Although the component values of equation (76) have units of stress, they do not necessarily give correct local stress values. The mathematical reason is that the divergence theorem, on which their derivation is based, has a meaning only for complete closed surface integrals. Physically, local stresses may be statically indeterminate and therefore not available at all from such a global calculation. However, their closed surface integral is guaranteed to have a physical meaning and to represent total force correctly. Of course, the surface may be closed only in a restricted sense. For example, in a periodic structure the surface need only embrace one period, the remainder of the structure being taken care of implicitly.

The expressions given above may be rewritten in terms of the normal and tangential components of flux density at each point on the surface. In two-dimensional problems the calculation of force requires the determination of the normal and tangential values of flux density at each point along the contour. Difficulties may be encountered as the result of numerical cancellation (the *subtraction of elephants* problem once again!) when Maxwell stress calculations are applied to finite element solutions. First of all, flux density components are obtained from potential solutions by differentiation, a process which commonly emphasizes errors. If the element discretization of the air region is relatively coarse, the error in evaluating the pointwise stresses may be considerable. Of course, this problem may be overcome by increasing the number of elements. Secondly, if the contour of integration actually passes through a node connecting several elements, the flux density is multi-valued, and no current CAD system incorporates any rational way of determining which value of flux density to choose, or whether to apply some form of smoothing or averaging. Finally, and probably most significantly, the numerical round-off problem already alluded to remains omnipresent.

If the tangential force is to be determined in a device such as an electric motor, as for the calculation of torque, then the tangential component of stress may not necessarily be directed continuously in the direction of rotation—any particular pole of the rotor body may be attracted both toward the stator magnetic pole opposite it and to the one behind it. The relative strengths of these two forces depend on the angle between the rotor and stator magnetic axes. However, the stress distribution evaluates both forces and the resultant is correctly the difference between them. As with the virtual work method, the useful force may be small compared to the individual, positively and negatively directed, force components. In some situations, substantial numerical errors may result.

In practical applications of the Maxwell stress approach to force calculation, it is advisable to evaluate the force using several contours and then to average the results. The deviation between results obtained for different contours will often serve to indicate the likely accuracy level achieved, even though strict upper and lower bounds are not available. If these cautionary notes convey the impression that the Maxwell stress approach has drawbacks, it is worth noting that it also has the advantage of being computationally cheap; it requires just one field solution, while the virtual work approach demands a minimum of two.

Current–Force Interactions

The force density of equation (67), used as the starting point for the Maxwell stress evaluation, may of course be employed directly in the calculation of electromechanical forces. Since (67) gives a body force density, the net force on a conductor, viewed as a rigid body, is

$$\mathbf{F} = \int_U \mathbf{J} \times \mathbf{B} \, dU. \tag{77}$$

This expression effectively involves integration only over the current-carrying regions and thus may involve little calculation. Its advantage is also its disadvantage: it cannot be employed to find the forces on magnetic objects which do not carry any current, and it is therefore of limited use. For example, in computing the torque of an electric machine, (77) is not useful, because it will not produce the forces tending to move the rotor, only the force exerted by the conductors on the slot walls. In general, this is a feeble force compared to the magnetic surface forces that move the rotor.

A specialized form of (77) has been used for over a century by electric machine designers. If all the current is confined to a filamentary conductor of length l, placed in a field of uniform flux density B, the general form of force in (77) reduces to

$$F = B \, i \, l, \tag{78}$$

where i is the conductor current. This expression will correctly give the forces, and hence the machine torque, if (and only if) the rotor is perfectly smooth and cylindrical; for only in that case are there no other forces acting on the rotor. It is curious to observe that removal of simplifying assumptions, as in introducing rotor slotting in this case, can sometimes actually make results worse!

Rate of Change of Inductance

Still another approach to computing electromechanical forces is based on finding the rate of change of inductance with displacement. This approach is closely related to the principle of virtual work and is particularly useful for devices whose desired end product is motion. The motion may be continuous, as in rotating or linear induction machines; or it may be transient, as in stepping motors or loudspeakers. The latter class includes a host of devices intended to supply kinetic energy on a pulsed basis, as for example in an electromagnetic hammer.

The following discussion is based on a ballistic linear actuator, the hammer of a door chime, as an example of a motional device in whose analysis forces are best calculated using the rate of change of inductance. The designer's aim here is to maximize the kinetic energy possessed by the hammer at the moment of striking its target, and conversely, to position the target so as to have it struck at the moment when the hammer possesses maximum kinetic energy. Fig. 15 illustrates the device in question, an axisymmetric solenoidal coil and a moving plunger. The plunger is attached to a nonmagnetic hammer, which is not shown in the drawing.

The ballistic actuator is forced to act by applying a voltage V to the terminals of its solenoid. When the voltage is suddenly applied, current begins to flow in the coil, and a magnetic force appears to accelerate the plunger, thereby endowing it with kinetic energy. Neglecting the coil resistance, the instantaneous power input to the coil is, in accordance with Faraday's law as given in equation (1),

$$i\,V = i\,\frac{d(Li)}{dt}. \tag{79}$$

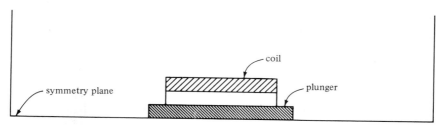

Figure 15. Ballistic actuator in its resting position. The plunger is centered within the coil structure.

This equation expresses a power balance: the input power must exactly equal the rate of increase of internal energy. Rewriting, (79) becomes

$$i\,V = i\,L\,\frac{di}{dt} + i^2\,\frac{dL}{dt}\,. \qquad (80)$$

The second term on the right-hand side results entirely from motion and represents the rate of change of stored energy. The rate of change of magnetically stored energy due to motion alone is obtainable from equation (23) as

$$\frac{dW_m}{dt} = \frac{1}{2}\,i^2\,\frac{dL}{dt}\,. \qquad (81)$$

The remaining energy must appear as the mechanical, i.e., kinetic, energy of the plunger. On the other hand, the kinetic energy is known to be expressible in terms of the mass m and the velocity v of the hammer and plunger as

$$W_k = \frac{1}{2}\,mv^2. \qquad (82)$$

Equating the kinetic energy to itself, there is obtained

$$\frac{1}{2}\,i^2\,\frac{dL}{dt} = \frac{1}{2}\,mv^2. \qquad (83)$$

Rewriting, using the chain rule of differentiation, there results

$$i^2\,\frac{dL}{dx}\,\frac{dx}{dt} = mv^2. \qquad (84)$$

This equation can finally be solved for the hammer velocity. Noting that

$$v = \frac{dx}{dt}\,, \qquad (85)$$

it immediately follows that v is given by

$$v = \frac{i^2}{m}\,\frac{dL}{dx}\,. \qquad (86)$$

Thus the velocity of the plunger can be determined from the change of inductance as the plunger moves from one position to the next. The computation required is thus to find the coil inductance L as a function of position, using an energy-based definition of inductance, then differentiating (or taking finite divided differences) and substituting in (86). The force acting on the plunger at any point in its travel may be determined subsequently by any one of the force calculation methods discussed above. Since the mass of the plunger is presumed to be known, its acceleration may also be determined.

The flux distribution for the plunger in its equilibrium (symmetric) position is shown in Fig. 15; Fig. 16 shows the flux lines when the plunger is displaced to one side. The field plots clearly demonstrate that the force is one tending to restore the plunger to move to the centered position.

For a plunger 1.25 times the length of the coil and an excitation of 300 ampere-turns, typical results for the centered position of Fig. 15 and for the off-center position shown in Fig. 16 are as follows:

Position:	Centered	Off-center
Magnetic energy (mJ)	1.555	0.771
Plunger kinetic energy (mJ)	1.418	0.634
Plunger velocity (m/s)	0.476	0.319
Force on plunger (mN)	0.0	57.0

The plunger is considered to be magnetic but nonconducting throughout this discussion. If the plunger is made of conductive material then the motion through the magnetic field would induce eddy currents which, in turn, would alter the device performance.

Local Field Values

The determination of local phenomena—field components, power densities, and particle forces—is an important part of the CAD art. Extracting

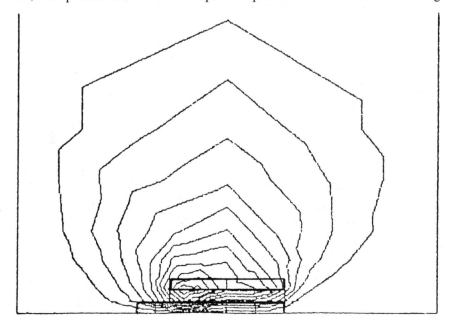

Figure 16. Ballistic actuator with plunger displaced to one side.

local results from solutions is therefore just as important as determination of inductances or forces. Of course, the extraction of local values requires somewhat different postprocessing operations.

Local phenomena rival global ones in importance because the electromagnetic device designer typically must satisfy two distinct forms of specifications: device performance and device feasibility. Both may involve either global quantities—stored energies, power levels, or total forces—or local values, such as flux densities, field uniformities, or particle trajectories. Devices used as *system components* interact with their environments through terminal parameters which reflect global performance specifications, although their feasibility limitations may be imposed by local internal phenomena. Thus a power reactor will be expected to have a particular inductance value, and an electric motor to furnish a specified torque; both designs will no doubt be constrained by local power loss densities or electric field gradients. In contrast to system components, *instruments* are often specified in terms of local behavior and limited by global considerations. An electron lens, for example, may have its performance prescribed in terms of aberration coefficients, which depend heavily on details of local field structure; its feasibility will be dependent on total power loss and material volume.

Local Field Components

Most present CAD systems employ the vector potential **A** to represent magnetic events, so that the determination of any field quantities must take this potential function as the point of departure. The most direct, and easily computed, quantity is of course the flux density, given in general by

$$\mathbf{B} = \text{curl } \mathbf{A}. \tag{12}$$

In two-dimensional cases where the vector potential is directed entirely into the plane of solution, only a single vector component appears; **A** seems to be a scalar quantity even though it is in reality a one-component vector. The general form (12) then assumes the special form

$$\mathbf{B} = \text{curl } (\mathbf{1}_z A). \tag{87}$$

$\mathbf{1}_z$ being the unit vector in the z direction. In terms of its Cartesian vector components, the flux density is thus

$$B_x = -\frac{\partial A_z}{\partial y}, \tag{88}$$

$$B_y = +\frac{\partial A_z}{\partial x}. \tag{89}$$

If required, the corresponding magnetic field values are found by first determining the local value of reluctivity, then multiplying. The reluctivity is of course a function of the entire flux density, not merely one component,

$$\nu = \nu\ (B_x, B_y),\tag{90}$$

so that to find the magnetic field **H**, determination of *both* components of **B** is necessary. This determination is generally done automatically by good CAD systems, and the user need not be explicitly aware of it, until the need arises to push the limits of the system. The field components are of course given by

$$H_x = -\ \nu\ \frac{\partial A_z}{\partial y}\tag{91}$$

and

$$H_y = +\ \nu\ \frac{\partial A_z}{\partial x}.\tag{92}$$

Comparable expressions are easily derived for components in other coordinate systems. In point of fact, most CAD work is conveniently done in either Cartesian or polar coordinates, as a matter of user preference. The finite element methods commonly used are substantially coordinate-independent, but users preferences tend to favor polar coordinates for the analysis of rotating devices, Cartesians for translational motion or symmetry.

Where a magnetic scalar potential function Ω is employed in solution, the vector quantity directly obtainable is the magnetic field **H**, not the flux density, for

$$\mathbf{H} = -\ \mathrm{grad}\ \Omega\tag{93}$$

whose components are clearly obtained by forming the partial derivatives

$$H_x = -\ \frac{\partial \Omega}{\partial x}\tag{94}$$

and

$$H_y = -\ \frac{\partial \Omega}{\partial y}.\tag{95}$$

To obtain the corresponding components of flux density **B**, H_x and H_y are multiplied by the local material permeability. In nonlinear materials, the permeability is a function of *all* components of **H**, as is evident from (90), so all components of **H** must be determined to find any component of **B**. While in principle the permeability may be a tensor quantity, few current CAD systems allow for tensor permeabilities to be taken into account automatically.

A Recording Head

An example of the need for local field values may be found in the examination of a magnetic recording head. The specification of a recording head must first and foremost be stated in terms of performance, that is, in terms of the quantity and quality of information it is actually capable of writing onto the recording medium. Secondary specifications may involve global values such as terminal impedance. The feasibility of a proposed design, however, is very much bound up with global values, such as the power loss in the device, and its associated temperature rise.

The primary point of interest in the analysis of a recording head is presumably determination of the fields and field gradients within the recording medium. The magnetization of the recording medium requires that the field gradient be high enough at the writing point, but it must be considerably lower elsewhere to avoid destroying information already recorded. Fig. 17 shows an outline drawing of a perpendicular recording head and the magnetic field produced by it, assuming the recording medium to be nonmagnetic. Ideally, the analysis should take care of both nonlinearity and hysteresis effects and might well include a reasonable model of the magnetic recording process itself. These requirements unfortunately surpass by a wide margin what currently available CAD systems are capable of doing. However, while awaiting the ultimate in software, the engineering analyst can still obtain a great deal of useful data from only a nonlinear analysis.

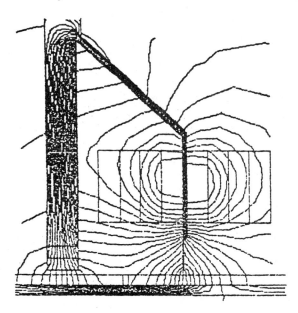

Figure 17. A perpendicular magnetic recording head. The recording medium is assumed to be nonmagnetic; it is backed by a magnetic material (bottom edge of drawing).

Similar requirements as to local field behavior arise in the design of large magnets for use in particle accelerators or nuclear magnetic resonance diagnostic machines. In these latter devices not only are the local field values important, but an even more crucial evaluative criterion is the *uniformity* of the field. It might be noted, however, that high field uniformity implies that the vector potential is required to vary linearly with position. This fact in turn implies that a very fine discretization of the air gap is not needed, at least for a first analysis!

Components in Local Coordinates

In many devices the local field values of importance are not the field or its components at a specific point, but rather along some track or contour. In the magnetic recording head example, the fields at specific points are certainly interesting; but most designers would consider a more global picture of greater interest and value than mere spot readings. Thus the first, essentially qualitative, representation of a solution desired by analysts is a flux plot, as in Fig. 17. But having examined the overall picture and found it satisfactory, the designer ordinarily wishes to move on to more quantitative data. In the present example, that probably means a curve such as Fig. 18, showing the perpendicular component of the magnetic field at the upper surface of the recording medium.

If the local values of field prove more or less satisfactory on examination, the designer may well attempt to modify the shape of the head, in order to improve the head performance as expressed by the curve in Fig. 18. The next calculation may very likely seek to determine the ampere-

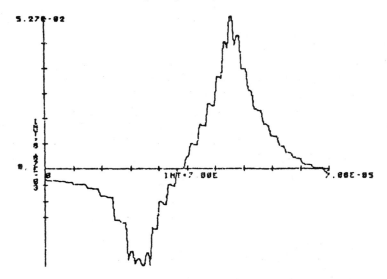

Figure 18. Vertical component of magnetic field under the recording head, at the top surface of the recording medium, plotted against longitudinal distance.

turn requirement of the iron member of the head itself. In other words, the designer may wish to evaluate the magnetomotive force M_{iron} given by

$$M_{iron} = \int_{iron} \mathbf{H} \cdot \mathbf{dl}. \qquad (96)$$

The contour of integration must follow an inverted U-shaped path, more or less as in Fig. 17. This path will not in general coincide with coordinate directions, and it will be necessary to resolve the vector $\mathbf{H}$ into components tangential to the path (parallel to the direction of the line segment $\mathbf{dl}$) and normal to it. Only the tangential component then enters the calculation in (96). This situation is similar to the component calculations encountered in the Maxwell stress expressions, equations (75) and (76), where the *normal* and *tangential* directions refer to the local orientation of the surface of integration or, in two-dimensional problems, the contour of integration.

In rotating electric machines, most contours of interest are either circular, or follow the complicated outline of some machine part. Probably the most commonly required field component plot in such equipment is the air gap flux density, for it has served generations of designers as a traditional working tool and has the immense advantage of familiarity. The radial component of flux density in particular gives both a qualitative and quantitative indication of the coupling across the machine air gap and may also be used in the calculation of the forces exerted on the shaft. The tangential component contributes to the calculation of the thrust forces.

Where field or potential values are required along a contour, for purposes of display or calculation, the Cartesian vector components can be resolved to yield the normal and tangential component values. Two reasonable ways of proceeding may be suggested: contours likely to be needed may be predefined in the CAD system, or the user may be given facilities for defining his own. Air-gap fluxes of machines, for example, are a very common requirement and justify inclusion of circular arcs as standard contours in general-purpose CAD systems; the same may be said of straight-line segments. Other, less common, shapes are probably better catered for by allowing them to be defined as needed.

Local field values to be displayed or further processed may be selected by location, or by value. Selection by location—choosing values at points or along contours—has already been discussed at some length. Selection by value typically implies searching. For example, the designer may inquire after the *highest value of flux density* and the place where it occurs; or, more broadly, he may wish to examine *all regions where the flux density exceeds 90% of its peak value.* Selection by value is often combined with selection by location, as in *highest flux density within the recording medium,* or *loss density above average value in the left half.*

To permit selection by location, interactive CAD systems may include facilities for the user to identify chosen sections of the problem model.

These sections may be two-dimensional portions (complete areas) within the model, such as might be needed in the calculation of the loss in a tooth; or one-dimensional portions, i.e. contours, along which the distribution of the field is required; or zero-dimensional pieces, i.e. points, at which the local value of the field (or its component in a given direction) might be needed. The geometric entities as well as the mathematical operations required are similar in principle, but different in details in each case.

To permit selection by value, control languages for postprocessing need to include suitable command verbs. This matter will be dealt with in some detail in connection with the structure of postprocessing programs themselves.

Determination of Terminal Voltage

Finding the motionally induced terminal voltage of rotating machines is a design requirement traditionally fulfilled by calculating the air-gap flux density. In CAD systems which solve for the vector potential, this approach is feasible, but usually bad. It does produce results, but does through tortuous computations what can be achieved simply and quickly, and with better accuracy, through a calculation based on flux linkages.

The induced motional voltage in a thin electric machine coil of n turns is easily calculated. Let the coil have one side placed at a particular point L in the $r-\theta$ plane, the other side at a point R. If the axial length of the machine is Z, then the flux linkages of this coil at any particular moment are given by

$$n\phi = nZ \ (A_R - A_L).\qquad (15)$$

The coil voltage is, as usual, given by Faraday's law,

$$e = -\frac{d}{dt}(n\phi).\qquad (1)$$

In a rotating machine, the points R and L will move along circular paths of fixed radii r_R and r_L, with their angular positions changing by an increment $d\theta$ in time dt. Thus, by the chain rule of differentiation,

$$e = -\frac{d\theta}{dt} \ \frac{d}{d\theta} \ (n\phi).\qquad (97)$$

The first factor on the right will of course be recognized immediately as the rotational speed of the machine.

Where coils of substantial cross-sectional area, and possibly nonuniform winding density, are employed, the above expression may be generalized by employing the broader definition (20) of flux linkages. A small current

i_c is imagined to flow in the coil, causing a current density J_c to exist in it. Integrating over S, the entire area occupied by the coil, (97) then becomes

$$e = -\frac{d\theta}{dt}\frac{d}{d\theta}\frac{\int_S A\, J_c\, dS}{i_c}. \tag{98}$$

The computational procedure implied by this expression is simple: the flux linkages are evaluated for two positions of the rotor. They are subtracted and divided by the angular displacement so as to form an approximation to the rate of change of flux linkages with angle. Multiplication by the rotational speed then yields the generated voltage directly.

In most practical cases the generated voltage can be found from a single field solution. It is not necessary to compute two solutions for two distinct positions of the rotor, for it usually suffices to leave the rotor and stator alone but to shift the position of the winding by one slot pitch. An exception to this general rule occurs when both rotor and stator are slotted, and the air gap is small. The generated voltage then includes a significant slot ripple component, which will be ignored if the rotor is effectively rotated by exactly one slot pitch. Recomputation of course involves extra work. There is consolation, however, in the fact that the conventional air-gap flux density method would not exhibit the slot ripple at all.

The traditional method of computing generated voltage examines the air-gap flux density, and in effect computes the generated voltage that would result if the coil were located in the air gap. The accuracy of this technique is lower than can be expected of direct flux linkage computation as above, and of course the amount of work is considerably larger. Unless there is some particular reason to the contrary, the air-gap flux method is therefore not to be recommended.

Fields in Axisymmetric Problems

Many problems encountered in magnetic devices have some form of axial symmetry. Solution in cylindrical coordinates actually produces true three-dimensional solutions in such cases. For example, the magnetic bearing and the ballistic actuator discussed above are actually axisymmetric devices.

When postprocessing operations are carried out on axisymmetric fields, care must be exercised on two counts. First, the common vector operators *grad, curl,* and *div* have forms which are different from those encountered in Cartesian coordinates. Secondly, most axisymmetric solver programs do not actually compute the vector potential, but a closely related potential-like function weighted by the radial coordinate r.

Differentiation and integration of vector quantities in cylindrical coordinates are not difficult operations, and their general forms can be found in standard books on electromagnetics. The classical axisymmetric problem has purely solenoidal current densities, with the vector potential **A** purely azimuthal, that is, possessing only a single component A_θ. In this rather special situation, the curl operation is simplified, so that the flux density components are now derived as

$$B_r = - \frac{\partial A_\theta}{\partial z} \tag{99}$$

and

$$B_z = \frac{1}{r} \frac{\partial}{\partial r}(r A_\theta). \tag{100}$$

The azimuthal component of **B**, which would have to be derived purely from r and z directed components of **A**, does not exist.

Most CAD systems do not solve for the magnetic vector potential directly because the axisymmetric Poisson and Laplace equations include a singularity at $r = 0$. The singularity arises from a multiplier of the form r^{-1} and is removed by modifying the potential by a similar but opposite multiplier. In the MagNet system, for example, the modified potential

$$U = \frac{A_\theta}{r} \tag{101}$$

is used. The components of **B** for this case are obtained by direct substitution into equations (99) and (100), as

$$B_r = - r \frac{\partial U}{\partial z} \tag{102}$$

and

$$B_z = 2U + r \frac{\partial U}{\partial r}. \tag{103}$$

In these expressions, it is interesting to note that r appears as a multiplier; thus U does not have to vanish along the axis of rotational symmetry in order for A_θ to vanish. That is to say, the modified potential U can be a regular function at the axis, which is precisely the reason for introducing it in the first place. The flux density components are regular also; in fact, at the axis itself, B_z is proportional to U, as can be seen readily from equation (102).

Some older CAD systems solve directly for the potential **A**, ignoring the singularity whenever it can be ignored and dealing with it in an ad hoc fashion wherever it cannot. In some special-purpose programs which solve for **A** directly, the regions modelled are in fact simply forbidden from including the axis of symmetry. There also exist other quite useful modified potentials different from U of equation (101). While the several ways of regularizing the differential equations all make solution feasible, it is important to be aware which modified potential is used before attempting to derive flux densities or any other quantities from it.

In axisymmetric problems, flux lines are not lines of constant A, but rather lines of constant rA. This fact easily follows from the general relationship

$$\phi = \oint \mathbf{A} \cdot \mathbf{dl}. \tag{13}$$

Consider a closed contour which proceeds radially outward from the axis of symmetry, then follows a circular arc at $r = constant, z = constant$ for an angular distance of θ_0, and finally returns to the starting point along a radial path. The flux ϕ_0 contained within this wedge-shaped contour may be evaluated by reference to (13). Since the vector potential $\mathbf{A}$ is everywhere purely azimuthal, no contribution accrues to the integral along the two radial sides, where $\mathbf{A}$ is orthogonal to the distance element $\mathbf{dl}$. Along the circular arc, $\mathbf{A}$ and $\mathbf{dl}$ are exactly parallel, so the integral evaluates to exactly $rA\theta_0$. Thus, in this particular case,

$$\phi_0 = rA \; \theta_0. \tag{104}$$

If the circular arc is displaced to some other point in the r-z plane in such a way as to keep ϕ_0 constant, then the new point must lie on the same flux line as the old, for if it did not, moving from one point to the other must have changed the flux linked. Flux lines in an axisymmetric problem are therefore lines of constant rA, not lines of constant A, and it is necessary to keep track of which are plotted in any given situation. Furthermore, if the modified potential U of equation (101) is employed, it follows that

$$\phi_0 = r^2 \; U \; \theta_0. \tag{105}$$

In other words, flux lines are lines of constant r^2U, and it is these which designers commonly wish to inspect. As a corollary, the requirement that the flux density $\mathbf{B}$ remain unchanged in the air gap of a magnet is equivalent to requiring that U remain constant. The modified potential U, in other words, is smoother than the vector potential $\mathbf{A}$ itself, hence computationally better behaved.

All the foregoing discussions for the calculation of terminal conditions, mechanical effects, and local values in the x-y plane remain valid in the axisymmetric case, provided due care is taken to use the correct forms of the differential operators, as well as the correct potential function.

Linear Time-Varying Problems

Many magnetic devices are initially analyzed on the assumption that the fields are essentially static, that is, no induced currents flow anywhere. For some devices refined analysis is subsequently required to determine the effect of eddy currents which actually do occur. Still other devices cannot work at all if no induced currents are allowed to exist. Most CAD systems therefore permit time-varying as well as static fields to be analyzed.

The postprocessing operations applicable to static field solutions, as described this far, are for the most part equally valid for time-dependent problems. However, there are several new quantities that a designer might require, which only have value in time-varying cases. Some of these will be reviewed in the following.

Time-Harmonic Analysis

In general, computer software for time-varying problems can be classified into time-domain and frequency-domain programs. Time-domain programs work by generating a sequence of solutions, one for each of a series of time values; frequency-domain programs solve for sinusoidal excitations at one or more fixed frequencies. This distinction may be familiar from circuit theory, where a corresponding classification of programs is possible. The time-domain approach is capable of dealing with arbitrary excitations applied to nonlinear devices, but its very generality is also its disadvantage: it generates extremely large data files at the cost of great quantities of computer time. In essence, it produces a movie film of the field behavior, one frame per time instant; all the frames need to be computed and stored. The frequency-domain technique is very compact and cheap, for the volume of data to be stored is just double that of a static solution, while the computing time required is greater than that for static solutions, but greater only by a modest amount. It is unfortunately applicable only to linear problems, because it is based on the premise that *all* time-dependent phenomena are sinusoidal, a premise satisfied only by linear systems.

Most present CAD systems make provision for analysis of time-harmonic phenomena in linear materials, but they do not provide equally extensive facilities for the general time-varying case. This restriction is not nearly so draconian as might seem, for many design specifications and many concepts on which system evaluation is based rely on notions of impedance, phase delay, and amplitude, notions which are strictly valid only for sinusoidally time-varying phenomena. The present discussion will therefore be confined to frequency-domain analysis.

As discussed in the chapter *The Potential Equations of Magnetics*, sinusoidally varying fields are conveniently described by combining the magnetic vector potential **A** and (sometimes) the electric scalar potential V. The vector potential is determined by solving the complex diffusion equation

$$\nabla^2 \mathbf{A} + j\omega\mu g\mathbf{A} = -\mu\mathbf{J}_O \tag{106}$$

which is obtained from the general time-varying case through replacement of the time derivative by the phase quadrature operator $j\omega$. The vector potential **A** is still assumed entirely z directed, but it is now a complex number rather than a pure real as was the case in static field problems. Here, as earlier, $\mathbf{J}_O$ represents the current density that would exist at extremely low frequencies, i.e., the current distribution that would obtain if the excitations all took place so slowly that no induced currents existed and all convection currents assumed a spatial distribution in accordance with material resistivities only. Briefly, though not totally accurately, $\mathbf{J}_O$ is sometimes referred to as the *dc distribution* of current density.

Of course, it must not be forgotten that the solution to Equation (106) really represents two solutions to two problems,

$$\nabla^2 \mathrm{Re}[\mathbf{A}] - \omega \mu g \, \mathrm{Im}[\mathbf{A}] = -\mu \, \mathrm{Re}[\mathbf{J}_O] \tag{107}$$

and

$$\nabla^2 \, \mathrm{Im}[\mathbf{A}] + \omega \mu g \, \mathrm{Re}[\mathbf{A}] = -\mu \, \mathrm{Im}[\mathbf{J}_O]. \tag{108}$$

The point here is that $\mathrm{Re}[\mathbf{A}]$, the solution of (107), represents the magnetic field at that moment in the a-c cycle when the real part of the driving current, $\mathrm{Re}[\mathbf{J}_O]$, reaches its peak and its imaginary part $\mathrm{Im}[\mathbf{J}_O]$ vanishes. This solution is altogether independent of the other solution $\mathrm{Im}[\mathbf{A}]$, which represents the field at the moment when the imaginary part $\mathrm{Im}[\mathbf{J}_O]$ reaches its peak and $\mathrm{Re}[\mathbf{J}_O]$ vanishes. At all other times, the real time-varying vector potential $\mathbf{A}(t)$, which might alternatively have been found by using a time-domain solution, can be determined by superposing suitably weighted components. Its value for any time t is given by

$$\mathbf{A}(t) = \mathrm{Re}[\mathbf{A}] \cos \omega t + \mathrm{Im}[\mathbf{A}] \sin \omega t. \tag{109}$$

By choosing a sequence of time values t, say at every $10°$ of the ac cycle, and evaluating the potential $\mathbf{A}(t)$ at each time instant in the sequence, a series of "snapshots" of the field can be generated. The two solutions $\mathrm{Re}[\mathbf{A}]$ and $\mathrm{Im}[\mathbf{A}]$ suffice for this purpose, no other field solution is required. Often, experienced analysts take the sequence of time values to lie at $90°$ intervals; in other words, they plot only $\mathrm{Re}[\mathbf{A}]$ and $\mathrm{Im}[\mathbf{A}]$ and do not bother with intermediate values. This abbreviated approach is suitable for anyone who has acquired considerable familiarity with time-harmonic field problems and is therefore capable of performing the required interpolation ("how **A** gets from $\mathrm{Re}[\mathbf{A}]$ to $\mathrm{Im}[\mathbf{A}]$") mentally. Most engineers are well advised to generate at least two or three intermediate plots as well, so as to gain a clearer appreciation of how the flux lines of one plot gradually merge and split in time to produce the time-quadrature picture.

As time goes on, magnetic flux lines in time-harmonic problems do not merely pulsate, they shift and twist about in space. The plotting of field patterns through time is therefore an important tool in learning how magnetic devices behave and how to make design improvements in them. As suggested above, one good way of plotting the fields is to produce a sequence of time "snapshots", movie frames as it were, which can be viewed in succession. Such sequences of snapshots can even be combined into motion pictures. Although actually making a motion picture is a very demanding technical task best turned over to professionals, the motion picture can sometimes communicate in seconds things which cannot be got across in minutes or hours of explanation. It is therefore of potentially very great value in management presentations or sales work. The device designer is well advised to keep in mind, while examining field plots, that novel devices or even design changes will need to be explained at some time motion pictures can communicate some things very effectively!

Current Densities

A quantity of frequent interest in time-harmonic fields is the density of induced (eddy) currents, and the density of total currents that results. In classical terminology, *eddy currents* are generally understood to be currents induced in conductive material which would not otherwise carry any current; the term *skin effect* is used to denote current density distribution in material that carries both excitation current ($\mathbf{J}_O$ in the above) and eddy currents. Here the term *induced currents* is employed to denote currents that arise from time-varying magnetic fields in both situations.

The induced currents that result from time variations in the magnetic field can always be obtained by combining Ohm's law,

$$\mathbf{J} = g\,\mathbf{E}, \tag{110}$$

with the general rule for electric fields

$$\mathbf{E} = -\frac{\partial \mathbf{A}}{\partial t} - \mathrm{grad}\ V. \tag{111}$$

Combining, and replacing the time derivative by the complex operator valid in the time-harmonic case,

$$\mathbf{J} = -j\omega g\,\mathbf{A} - g\,\mathrm{grad}\ V. \tag{112}$$

Here $\mathbf{J}$ is the total current density, attributable to all the mechanisms that make current flow. However, the two terms on the right-hand side may be identified separately, as the dc current distribution

$$\mathbf{J}_O = -g\,\mathrm{grad}\ V \tag{113}$$

and the *induced current density*

$$\mathbf{J}_i = -g\,\omega\mathbf{A}. \tag{114}$$

It should be noted that the general equation (112) relies only on the definition of **A** as a vector whose curl produces the flux density **B**; it is therefore quite independent of the choice of gauge.

Considerable interest attaches to plotting the eddy current densities that occur in conductive parts of various magnetic devices. In the two-dimensional cases discussed here for the most part, **A** is z directed or (in axisymmetric problems) θ directed, so the induced current density $\mathbf{J}_i$ must therefore be z directed or θ directed also. Plotting is therefore not a fundamental problem—there is still only one component of **J** for a single-component **A**—but it is slightly more complicated because the form of display is less well established conventionally. While it is perfectly possible to draw contours of equal magnitude of current density, such lines do not always convey the desired information. A better choice, in CAD systems which permit it, is to produce a *zone plot*. In such a plot, the space to be plotted is divided into a set of *zones*, so that one zone includes all that portion of space in which the current density lies between its maximum value and (say) 90% of maximum; another zone in which the current density lies between 90% and 80% of maximum, and so on down to zero. Plotting is then done by filling in the highest-density zone with a very bright color, the next zone with a less bright color, . . . , and the zero-density zone with a dark color. People generally find it easy to associate bright colors with high densities, dark colors with low ones, and can therefore interpret zone plots rather easily.

Equiphase plots of current density values can sometimes be quite illuminating. These will be considered more fully later, in connection with equiphase plots of other phasor quantities such as fields and the Poynting vector.

Power Flow and the Poynting Vector

The Poynting vector is a quantity generally associated with the flow of energy in electromagnetic devices and used to measure power transferred. It is established by considering the stored energy and the power dissipated in a device and relating them through the law of energy conservation.

Let W_m be the magnetically stored energy in a closed region of space U. If the region contains magnetically linear materials, as it must if time-harmonic analysis is to be useful, then this energy is given by

$$W_m = \frac{1}{2} \int_U \mathbf{B} \cdot \mathbf{H} \, dU, \tag{115}$$

while the electrically stored energy W_e in the same region is

$$W_e = \frac{1}{2} \int_U \mathbf{E} \cdot \mathbf{D} \, dU. \tag{116}$$

The power P_e dissipated in the region U is similarly given by

$$P_e = \int_U \mathbf{E} \cdot \mathbf{J} \, dU. \tag{117}$$

Now the law of conservation of energy clearly requires that the input power P into the region U must equal the dissipation plus the rate of increase of stored energy. That is,

$$P = P_e + \frac{\partial}{\partial t} (W_m + W_e). \tag{118}$$

Substituting the explicit expressions for P_e, W_m, and W_e from above, the input power turns out to be

$$P = \int_U \mathbf{E} \cdot \left[\frac{\partial \mathbf{D}}{\partial t} + \mathbf{J} \right] dU + \int_U \mathbf{H} \cdot \frac{\partial \mathbf{B}}{\partial t} \, dU. \tag{119}$$

But the factor in parentheses, and the time derivative of $\mathbf{B}$, are readily recognized as right members of the Maxwell curl equations. Substituting, there finally results

$$P = \int_U \mathbf{E} \cdot \text{curl } \mathbf{H} \, dU - \int_U \mathbf{H} \cdot \text{curl } \mathbf{E} \, dU. \tag{120}$$

A well-known vector identity permits rewriting the latter in terms of the cross-product of the vectors $\mathbf{E}$ and $\mathbf{H}$ as

$$P = \int_U \text{div} (\mathbf{E} \times \mathbf{H}) \, dU, \tag{121}$$

which may be converted, making use of the divergence theorem, into the closed surface integral over the bounding surface ∂U of the region U:

$$P = \oint_U (\mathbf{E} \times \mathbf{H}) \cdot \mathbf{dS}. \tag{122}$$

The vector in parentheses is usually known as *Poynting's vector* and denoted by $\mathbf{S}$,

$$\mathbf{S} = \mathbf{E} \times \mathbf{H}. \tag{123}$$

The Poynting vector is conventionally thought to measure local power flow, much as W_e and W_m are held to measure local stored energy densities. It is often plotted and viewed as a quantity indicative of the performance of a device, especially in so far as portions of a device may be concerned. If some part of a device hardly matters in power flow or energy transformation, should it be there at all?

The Poynting vector may be written in terms of the vector potential **A**, so as to facilitate its evaluation in CAD systems. Substituting from (12) and (111) for the magnetic and electric quantities respectively, equation (123) assumes the form

$$\mathbf{S} = -j(\omega/\mu)\,\mathbf{A} \times \operatorname{curl}\mathbf{A} - (1/\mu)\,\operatorname{grad} V \times \operatorname{curl}\mathbf{A}, \qquad (124)$$

which only involves the potentials normally available from field solutions.

The Poynting vector is generally a two-component quantity even in problems where only a single component of vector potential exists. It is difficult to plot in a graphic display, because there is no conventional form of representation for time-varying multicomponent vectors. One fashion in which the Poynting vector, and indeed other planar vectors such as **B** or **H**, can be displayed is in two independent plots, showing phase and magnitude.

Equal-magnitude plots of vector quantities are often best arranged as zone plots, while equiphase contours are best left as line contours. The reasons are almost solely those of ease in visual perception and interpretation. Energy transport (power flow) in a time-harmonic system is essentially a wave phenomenon: a packet of energy is sent off from its source and allowed to propagate toward its eventual sink, possibly attenuating through dissipation as it travels. In one period, any point in the region U executes one full cycle of time events. Conversely, the wave or energy packet propagates during one period precisely as far as it is necessary to go to find another space point where electrical events are 360° lagging in phase. In other words, energy travels in one period from one equiphase contour to the next one 360° removed. Plotting the equiphase contours thus makes evident the flow of energy through the device, and thereby exhibits to the experienced designer whether and how device performance might be improved.

Exploiting Superposition

In time-harmonic problems the assumption of magnetic linearity is essential, for in its absence the notions of sinusoidal phasor analysis become invalid. However, if linearity is assumed, then superposition may be used, permitting simplification of many problems.

Where a magnetically linear device has altogether N distinct excitation ports—windings, conductors, or other current-carrying entities—at most a total of N distinct field solutions is required at any one frequency to cover *all* possible cases. To choose a specific example, consider a transformer with N windings, in which eddy current losses may occur in the laminations or in other conductive materials. Now any possible excitation state of this transformer may be described by specifying all its winding currents, all of which jointly may be viewed as the driving current density $\mathbf{J}_O$ of equation (106). If $\mathbf{J}_{Ok}$ is the current density distribution caused by a *unit* current in winding k, then any set of winding currents $\{i_k, k = 1, \ldots, N\}$ will cause the excitation current distribution in the transformer to have the value

$$\mathbf{J}_O = \sum_{k=1}^{N} i_k \, \mathbf{J}_{Ok}(x,y). \tag{125}$$

Now a solution may be sought to equation (106) in the presence of only one of the winding currents. That is to say, all but one of the i_k may be set to zero, that one being given unity value. Equation (106) then assumes the form

$$\nabla^2 \mathbf{A}_k + j\omega \mu g \, \mathbf{A}_k = -\mu \, \mathbf{J}_{Ok}. \tag{126}$$

In the presence of many different currents i_k, the solutions $\mathbf{A}_k$, each valid for a single unit current, superpose. For the combined current $\mathbf{J}_O$ of (125), the relevant differential equation is thus (106), and its solution is immediately obtained as

$$\mathbf{A} = \sum_{k=1}^{N} i_k \, \mathbf{A}_k. \tag{127}$$

There is never any need to solve more than N field problems, for once the *N basis solutions* of (126) have been generated, any and all other solutions can be obtained as in (127). It should be noted that any eddy currents which may exist in conductive parts of the device are included in the solutions of (126), and therefore also of (127).

When solutions are constructed by superposition, it is worth noting that the solver phase of the CAD system may at times not be used to produce any actual solution at all. It may be employed solely as a generator of basis solutions, with the composite solutions, as in equation (127), created by a postprocessing program. Considerable computation time may be saved by this approach since the fastest available solvers operate in approximately $O(N^2)$ time whereas scaling and adding operations occur in $O(N)$ time.

Electrostatic Calculations

The operations and solution systems described earlier in this book have all referred to magnetic systems. However, many problems of an electrostatic nature are also described by either the Laplace or Poisson equation. Such electric field problems can be solved by using magnetostatic packages provided the user can determine the appropriate equivalences. To illustrate, a simple capacitor with mixed dielectrics may be considered.

A Parallel-Plate Capacitor

A parallel-plate capacitor, imagined to extend infinitely into the paper, is illustrated in Fig. 19. Initially, the model is defined with a dielectric of air and only one of the possible symmetries in the problem is used. The equation to be solved is a scalar Laplace equation in which the scalar potential is the voltage in the system. The electric field $\mathbf{E}$ is related to the electric scalar potential by

$$\mathbf{E} = - \text{grad } V, \tag{128}$$

a special case of (111), with no time variations.

When an equivalence is drawn with magnetostatics problems, the scalar potential V in the electrostatic model is substituted for the vector potential $\mathbf{A}$ of magnetostatics. The magnetostatic Poisson equation for nonlinear materials may be written

$$\text{curl } \nu \text{ curl } \mathbf{A} = \mathbf{J}, \tag{129}$$

whose mixed-dielectric electrostatic equivalent is

$$\text{div } \epsilon \text{ grad } V = -\rho. \tag{130}$$

In a two-dimensional Cartesian coordinate system these two equations reduce to identical form, when written out in detail. Hence an equivalence may be drawn between the current and charge densities, the vector and scalar potentials, as well as the material properties (reluctivity and permittivity). The latter equivalence is obvious but must be treated with caution if modelling of material property curves is required, for most magnetic material modelling programs allow permeability, rather than reluctivity, to be specified by the user. That is, the user needs to specify an $\mathbf{E}$ against $\mathbf{D}$ curve.

Once the analogous quantities have been identified, solution of the electrostatic problem proceeds exactly as if the problem were magnetostatic.

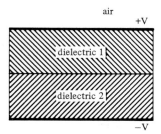

Figure 19. A simple parallel-plate capacitor.

For the capacitor of Fig. 19, an equipotential plot is shown in Fig. 20. Both dielectric slabs are assumed to be air in this case. The analogy with magnetics problems is immediately apparent on inspection of the solution. Although the finite element subdivision employed for the solution is not explicitly shown, it can be seen from Fig. 20 that the elements are generally quite large. An analogous solution, for the same geometric shapes but two different dielectrics, appears in Fig. 21. The lower portion of the dielectric space in this case has a relative permittivity of 20. The electric field can be seen to be considerably lower in the dielectric than it is in air.

The capacitance of multi-dielectric systems is probably the quantity of major interest, analogously to inductance in magnetics problems. It may be calculated by using substantially similar techniques, as will now be discussed in some detail.

Determination of Capacitance

The capacitance of an electrostatic device may be determined in either of two ways, which are analogous to the ways of computing inductance in

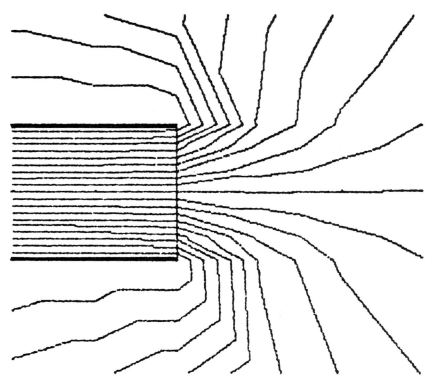

Figure 20. Electric field of the capacitor, with all dielectric material having the permittivity of air.

magnetics problems. One approach requires a calculation of the stored energy, the other a determination of the total electric flux. The capacitance C may be defined in two corresponding ways, either in terms of the electrically stored energy W,

$$W = \frac{1}{2} CV_0^2, \tag{131}$$

or as the ratio of charge to potential difference V_0 between the electrodes,

$$C = \frac{q}{V_0}. \tag{132}$$

The total electric flux is, by definition, equal to the total charge q. To determine it, one or the other electrode may be encased in a closed surface S, to which Gauss' law is then applied:

$$q = \oint_S \mathbf{D} \cdot \mathbf{dS}. \tag{133}$$

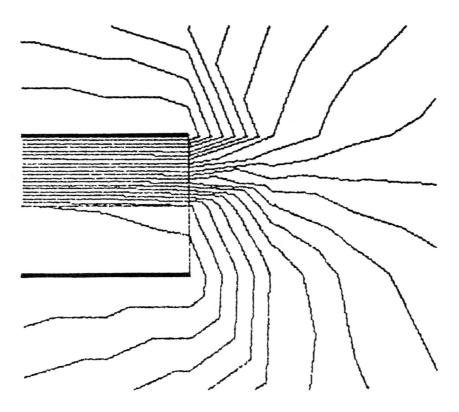

Figure 21. Electric field of the capacitor with mixed dielectrics.

The capacitance C is then found by combining (132) and (133) to yield

$$C = \frac{\oint_S \epsilon \text{ grad } V \cdot \mathbf{dS}}{V_0}, \tag{134}$$

where (128) has been employed to eliminate the flux density and to substitute the computed quantity V. It should be noted that any surface S which completely encloses the electrode is equivalent to any other in principle, since Gauss' law is equally applicable in all cases. In two dimensions, the surface integral of course reduces to a closed contour integral in the x-y plane.

Finding the capacitance via the energy route involves use of equation (116) to determine the stored electric energy, then equation (131) to determine C. Combining the two, there is obtained

$$C = \frac{\int_U E^2 \, dU}{V_0^2}. \tag{135}$$

Because finite element methods generally are energy-based methods, (135) usually provides the more accurate basis for calculation.

To illustrate, the capacitance was determined for a capacitor having a plate width of 0.1 m and a plate separation of 0.01 m. The gap is half filled with air and half filled with a dielectric of given relative permittivity, taken as either unity (to correspond with Fig. 20) or 20. Both an energy calculation and a contour integration were used. In the latter case, two contours were chosen, both the full width of the model; the first passed through the upper half of the air gap, the second through the lower half. Thus, for the dielectric model, the second contour passed through the dielectric layer. Values for all three calculations are given in Table 5. These results may, of course, be checked roughly by the usual simple formula for a parallel-plate capacitor, provided that the fringing at edges is neglected. This calculation gives a value of 88.5 pF for the air capacitance and 168.6 pF when the dielectric is present. The actual values, when fringing is considered, should of course be higher.

Characteristic Impedance and Phase Velocity

A two-dimensional problem of some interest is that of a parallel-strip transmission line similar to the mixed-dielectric capacitor described above.

Table 5. Capacitance (pF) of parallel-plate capacitor

Permittivity	Contour integral		Stored energy
	1	2	
1.0	99.34	98.81	110.94
20.0	181.95	194.34	196.12

If a quasi-TEM wave is assumed to propagate along such a line, the wave impedance and phase velocity may be determined directly from the two capacitance values computed above. The characteristic impedance of any transmission line is given by

$$Z_0 = \left[\frac{L}{C} \right]^{1/2},$$ (136)

while its phase velocity, for quasi-TEM waves, is

$$v = \frac{1}{(LC)^{1/2}}.$$ (137)

When the dielectric material is air throughout, the phase velocity must be equal to the velocity c of light. If C_0 is the capacitance in that case, then

$$c = \frac{1}{(LC_0)^{1/2}}$$ (138)

and hence the line inductance is

$$L = \frac{1}{c^2 C_0}.$$ (139)

Since both dielectrics have the permeability of free space, the inductance value is unaffected by the presence of the dielectric. Thus, the characteristic impedance of (136) becomes

$$Z_0 = \frac{1}{(c^2 C_0 C)^{1/2}}$$ (140)

and the phase velocity

$$v = \frac{C_0}{C^{1/2} c}.$$ (141)

For the device described above the characteristic impedance is then 22.6 ohms.

Effects of Numerical Approximations

Finite element methods form the mathematical basis for nearly all current general-purpose CAD systems intended for magnetics problems. In post-processing, the various mathematical operations—differentiation, integration, multiplication, and whatever else may be required—cannot be performed on the exact fields themselves, but must use the finite element approximations computed by solver programs instead. The precision of results, and even their interpretation, may be affected by the numerical approximations involved, and some brief consideration of the nature of these approximations may not be out of place here.

Finite Element Approximations

Finite element methods invariably produce polynomial approximations to field solutions. Over each finite element, the potential field is computed as a polynomial expression in the coordinate quantities, whose degree is usually fairly low. The approximations are so constructed that the potential solution is continuous everywhere; but its derivatives are not continuous across the element edges. If the potential is imagined to describe a surface above the x-y plane, the surface is described by a polynomial over each of a set of patches. At the patch edges, the pieces of surface join so there are no "holes", but the joints are not necessarily smooth; there may be creases. Thus, if elements of order N are used, the potential is given by polynomials of order N within each element, but it is C^0 continuous globally. Arithmetic and analytic operations in postprocessing operate directly on the polynomials, and the operations may on occasion trespass on the limits of the possible.

One immediate result of piecewise continuity is that the flux density distribution along a specified contour through a problem model will not be smooth. The flux density plots shown at various points throughout this book make this phenomenon quite evident. It is of course perfectly possible to beautify the results by polynomial (or other) interpolative smoothing. In general, the plots shown in this book have not been subjected to any cosmetic improvement, in the belief that raggedness in results, should it occur, conveys information about the reliability of results, and it would be a mistake to destroy this additional information.

The postprocessing operations outlined in this chapter include a broad variety of mathematical operations. Not all of these are permissible on piecewise polynomial functions in all cases; and not all produce piecewise polynomial results from piecewise polynomial arguments. At least the following elementary mathematical operations are included in the list of requirements:

Arithmetic: addition and subtraction of potential or vector fields, multiplication of fields by scalars.

Arithmetic: multiplication and division of potential (scalar) field quantities.

Vector Operations: cross and dot products of vector fields.

Functional Operations: absolute values, square roots, trigonometric functions, and other general functions of potential fields.

Differential Calculus: gradients of potential fields; curls and divergences of vector fields.

Integral Calculus: line and surface integrals over defined geometric regions.

These will be briefly examined in the following, with a view particularly to the accuracy to be expected in practical cases.

Arithmetic Operations

The elementary arithmetic operations fortunately do not affect the nature of finite element approximations. When piecewise polynomial functions of degree N or lower are added, the result is always a field of piecewise polynomials of similar degree. Thus the elementary arithmetic operations—addition, subtraction, and multiplication by a scalar—may be performed without effect on the nature of numerical approximation. As always, however, a cautionary note must be sounded regarding subtraction: small differences of large numbers may exhibit round-off error accumulation to a considerable extent. Wherever possible, it is thus preferable to formulate problems in such a way as to avoid numerical formation of small differences.

Multiplication of fields invariably produces results which are piecewise polynomial, but not of the same degree as the arguments. For example, multiplication of a first-degree polynomial field by itself, thus creating, say $v^2(x,y)$ from $v(x,y)$, produces a piecewise polynomial field of degree 2. No existing CAD system can handle the growth in polynomial order that results from sequences of repeated multiplications, and the only practical approach is to approximate the high-order polynomials by others of lower order.

Division of piecewise polynomial fields can only yield piecewise polynomials in rare cases. In general, the result is a piecewise rational fraction, continuous but not necessarily bounded. Again, no CAD system is capable of handling such functions, and the usual technique is to approximate them by piecewise polynomials in a least-squares sense. It is clear that some of the fundamental properties of the functions being approximated might be lost in the process. For example, the quotient $u(x,y)/v(x,y)$ will be unbounded if $v(x,y)$ vanishes anywhere in the region of interest; but an approximation to the quotient cannot be unbounded if it is to be expressible in terms of polynomials. The division operation therefore requires particular caution.

Operations on vector quantities are not inherently different from operations on scalars. The inner product of two vector fields, after all, is expressible as the combination of products of scalars, the vector components; thus, any problems likely to arise with scalar fields are likely to recur when vectors are considered. However, it must be noted that troubles arising from the division operation are entirely absent, since division of vectors is undefined.

Differentiation and Integration

Finite element methods produce piecewise polynomial approximations continuous in function value, but not necessarily continuous in any derivatives. Consequently, conventional finite element approximations are strictly differentiable only once. Within elements, the approximation of degree N is reduced to an approximation of degree $N-1$ by differentiation; but at element edges, differentiation destroys continuity. Indeed it cannot do otherwise, for the boundary conditions of electromagnetics dictate that at least some derivatives must be discontinuous at interfaces between dissimilar materials.

Differentiation operations on field quantities should generally be avoided wherever possible. This is not to say they should never be attempted; that would be foolish. However, it should be kept in mind that differentiation is always an error-emphasizing process, and that second or higher derivatives should be avoided. Avoidance is often possible by reformulation and integration by parts. For example, Green's second identity,

$$\int_U u \, \nabla^2 v \, dU = \oint u \, \frac{\partial v}{\partial n} \, ds - \int_U \nabla u \cdot \nabla v \, dU, \qquad (142)$$

is widely employed in field analysis, to eliminate the second derivative on the left in favor of more derivatives, but of lower order, on the right. On occasion, second and even higher derivatives can be found with reasonable accuracy, but system users invariably do so at their own peril.

Integration is generally an error-attenuating process, in which fluctuations are averaged out and results smoothed. It is therefore a safe process to employ with any function, and multiple integrals are safer still.

Difficulties do very occasionally arise with numerical integration, especially contour integration, where the region of integration includes extremely sharp fluctuations in the integrand. All numerical integration schemes work by sampling the integrand at a set of cleverly selected points, and combining the samples in some way. If the region of integration includes some important feature which only occupies a very small portion of the region, it is possible for the integration process to miss the feature altogether when performing the sampling. For example, a small hole in a two-dimensional region may be missed. This difficulty probably arises most often in magnetics problems in connection with contour integrals employed to calculate magnetomotive forces. If the contour is very long and crosses a very small air gap, it is just possible to have the integration procedure miss the air gap altogether!

Axisymmetric and Nonlinear Problems

Axisymmetric problems may give rise to numerical difficulty because the potentials are multiplied or divided in many operations by the radial distance from the axis of symmetry. As an example, the flux density $\mathbf{B}$ in such a problem is derived from the product $r\mathbf{A}$ rather than from the original vector potential field $\mathbf{A}$ for which the solution may have been produced. The problem is worse if a modified potential, such as $\mathbf{A}/r$, was used for the solution. The vector potential itself may be modelled by using first-order elements; thus it varies linearly. The coordinate r certainly varies linearly over the problem and the product term $r\mathbf{A}$ is developed by multiplying these two. What is required is a polynomial multiplication, and the true result of two first-order polynomials multiplied should be a second-order polynomial. As with the arithmetic operations already discussed, such high-order forms of data are very rarely used. The more common procedure is to approximate them in turn by polynomials of order equal to that employed by the finite elements in the first place.

Similar approximation problems arise with any nonlinear operator such as the square root, the trigonometric functions, or exponentiation. In each case the true result will not necessarily be expressible in terms of the original finite element approximating functions. The extent of the error caused by ignoring these difficulties and approximating all operators as producing piecewise polynomials will depend on the discretization of the original problem and the rate at which the field varies at any particular point.

Postprocessing Systems for Magnetics

When a solver program terminates its run, it leaves in a computer file a digital version of the magnetic field as calculated. The postprocessor is expected to perform any further manipulations—which may be many, and which may involve far from trivial amounts of computation—and must therefore be endowed with the capacity for performing them. Some of the varied techniques available for the extraction of engineering information from mathematical solutions were examined at length in the previous chapter. However, no attempt was made there to provide an exhaustive list of all the operations that might be needed; the stress was rather on results which are common to many designers' requirements. The present chapter is directed the other way: it seeks to ask *how* rather than *why*, and therefore to examine the implementation of a postprocessing system capable of meeting the requirements implicitly defined by its predecessor.

Postprocessor Structure and Use

The manipulations required in postprocessing, as reflected in the topics covered by the previous chapter, may well seem obvious to anyone thoroughly familiar with electromagnetic fields. But such indeed is the nature of most computing, not only postprocessing: the precise specification and exact execution of operations which, taken one at a time, are all but trivial. The great difficulties lie in managing complexity, "getting it all right", not in the inherent difficulty of the steps themselves.

A reasonable place to start a discussion of postprocessing systems is by analyzing the basic operations used in the preceding chapter, with a view to determining what actions a processing program must actually be capable of performing.

Manipulation of Fields

A good idea of the manipulative operations required in dealing with two-dimensional fields can be gained from an examination of the simulations of the inductor and the transformer given above. Their equivalent circuit parameters can be derived in several ways, each comprising a sequence of operations which produces the needed component values. The operations are essentially those of the vector calculus and may be listed as follows:

1. Finding the curl of a vector, for calculating flux density.
2. Surface integration over all or part of the model.
3. Line integration over a specified contour within the model.
4. Accessing the loss and magnetization curves.

These operations may seem to require no more than the elementary vector operations implemented on a computer; but they actually involve more than that. Indeed, they might be broadly classified in three categories: mathematics, data access, and result display. The first is obvious from the equations themselves; the second is necessary for specification of geometric features such as integration contours (requiring access to the geometric data base) or of material properties (requiring access to the material files). The third is a requirement of human communications, for computing a result is not much use if it cannot be presented in a useful way.

It is perhaps instructive to examine the particular problem of determining losses in a transformer. If the solution to the problem has been determined using a magnetic vector potential approach, then the flux density $\mathbf{B}$ is derivable from the potential $\mathbf{A}$. In two-dimensional problems where $\mathbf{A}$ is purely z directed, $\mathbf{B}$ is a vector having two components B_x and B_y. Thus, to obtain the total flux density required for the calculation of losses, its magnitude must be found. The steps to obtain this value can be described in terms of vector algebra. First, the components of flux density are found from the curl of the potential,

$$\mathbf{B} = \text{curl } \mathbf{A}. \tag{1}$$

Next, the two flux density components are extracted by way of a scalar product operation,

$$B_x = \mathbf{1}_x \cdot \mathbf{B}, \tag{2}$$

and

$$B_y = \mathbf{1}_y \cdot \mathbf{B}, \tag{3}$$

where $\mathbf{1}_x$ and $\mathbf{1}_y$ are unit vectors in the x and y directions. The magnitude B is then found by forming squares through multiplication,

$$B_x^2 = B_x \, B_x, \tag{4}$$

then adding and extracting the square root,

$$B = \sqrt{B_x^2 + B_y^2}. \tag{5}$$

The local loss density w is finally obtained by accessing the materials data files, looking up the function appropriate to the material m at hand, and evaluating it for the correct flux density B,

$$w = w_m(B). \tag{6}$$

The total loss is then obtained by integrating over the space S_m occupied by laminations made of material m,

$$W = \int_{S_m} w \, dS. \tag{7}$$

Even for such a relatively simple operation as loss estimation, a whole set of separate manipulations is obviously required. Note that this set includes both vector and scalar operations.

The field that a postprocessor can expect from the solver may be either vector or scalar in nature. However, both kinds of field can easily occur in postprocessing, since the gradient operator (and certain other operators) creates vector fields out of scalar ones, while the divergence operator reduces vectors to scalars. Accordingly, both vector and scalar operations must be provided. The functions that should be available within the post-processor for two-dimensional data may therefore be summarized by the following:

Vector Analysis: curl
                     div
                     grad
                     contour integral
                     surface integral
Vector Algebra: cross product
                     dot product
                     addition
                     subtraction
Scalar Algebra: addition
                     subtraction
                     multiplication
                     division

plus, of course, the data access and display functions appropriate to two dimensions.

Manipulation of Local Data

Various frequently encountered postprocessing operations require integration or differentiation, or indeed display of values, along contours rather than areas. As a simple example, the calculation of magnetomotive force difference between two points P and Q, say, requires evaluation of

$$F_{PQ} = \int_P^Q \mathbf{H} \cdot \mathbf{dl}. \tag{8}$$

This computation will require similar, but not identical, steps to the fore-going. It will again begin with an evaluation of the flux density,

$$\mathbf{B} = \text{curl } \mathbf{A},\tag{1}$$

continue with a function determination to find the value of reluctivity,

$$\nu = \nu\,(B),\tag{9}$$

then proceed with determination of the magnetic field $\mathbf{H}$ through that material property,

$$\mathbf{H} = \nu\,\mathbf{B}.\tag{10}$$

The magnetic field is of course a vector quantity. To integrate it, the contour from P to Q must be specified. Once the contour is known, the component of $\mathbf{H}$ tangential to the contour can be found at all points of the contour,

$$H_t = \mathbf{1_t} \cdot \mathbf{H},\tag{11}$$

where $\mathbf{1_t}$ is the unit vector in the direction tangential to the contour. Integration then takes the obvious course,

$$F_{PQ} = \int_P^Q H_t\,dl.\tag{12}$$

In this sequence, a two-dimensional field is dealt with up to the moment the tangential component of $\mathbf{H}$ is required. The term *tangential* only makes sense with reference to a particular contour and therefore has no meaning until the contour is defined.

An alternative procedure may be employed for determining magneto-motive force, which does not require explicit calculation of the field $\mathbf{H}$. Equation (12) can be rewritten as

$$F_{PQ} = \int_P^Q \nu\,\frac{\partial A}{\partial n}\,dl,\tag{13}$$

provided the magnetic material is isotropic (though not necessarily linear). In this case, differentiation in the direction normal to the contour is required. Again, this requirement is meaningful only with reference to a specified contour.

Operations along contours thus demand the ability to specify the contours themselves, as well as the ability to perform a set of operations more limited than those required on two-dimensional field regions, but not quite the same. They are

Analysis: directional differentiation
            contour integral

Algebra: addition
              subtraction
              multiplication
              division

and the associated operations for defining contours and for displaying results.

Of course, on quite a few occasions purely local values are required: the flux density or magnetic field at a point, for example, or perhaps the x directed component of the Poynting vector. For manipulation of data defined only at a point, little is needed beyond the operations available on an ordinary hand calculator:

Algebra: addition
              subtraction
              multiplication
              division

Once again, it is necessary to have facilities to identify the point at which the values are desired.

It might be noted that the manipulations indicated in the examples above, and in the previous chapter, refer to the truly continuous field rather than to a discrete approximation. Where equivalent alternative formulations exist, they are on occasion equivalent in the continuum, but one may be preferred over another in the discrete case on grounds of accuracy.

Calculator Structure

So far, the discussion has concerned the possible operations which might be required in a useful postprocessor. The set given above is by no means an exhaustive list as the reader might be aware from the previous chapter and it is likely that a useful implementation will contain operations beyond those listed above. However, any extra operations may still be classified within the headings described above, that is, input data being two-, one- or zero-dimensional, output data being two-, one- or zero-dimensional, and operations being vector or scalar. Consequently, the following discussion will be concerned with the fundamental structures for handling each data type: a two-dimensional, or *field*, calculator, a one-dimensional, or *curve*, calculator, and a zero-dimensional, or *numeric*, calculator.

For simplicity of implementation and use, these three types of data and their appropriate operations imply that three distinctly separate postprocessors, or calculators, are needed.

Graphic Output

Engineers, and electrical engineers in particular, are a very graphics-oriented profession. Engineering reports, work proposals, or management presentations may on occasion include no mathematics, but they very rarely get by without any graphs, drawings, plots, or diagrams. In engineering parlance, to *understand* something means, as often as not, to *be able to draw pictures* of it. Not surprisingly, postprocessing begins for many designers with the drawing of flux plots and continues through sequences of graphic representations of everything from Maxwell stresses to permeability values.

Graphic displays are intended primarily to communicate facts or ideas to people and should therefore be regarded as a particular form of human language. Like other human languages, graphics must be learned, and a comfortable facility with it must be acquired before it can be used effectively. Various specific forms of representation—the visual vocabulary of engineering graphics, as it were—are conventionally accepted, and CAD systems must adhere, at least in the main, to the conventional representations if they are to find acceptance by the engineering community. This constraint is both a blessing and a curse; widely understood conventions make learning and using CAD systems easy, but at the same time perpetuate their limitations, thereby discouraging system designers from improving on the methods of the pencil-and-paper era.

The entities currently dealt with in magnetics CAD are scalar or vector quantities, defined over regions which may be two-dimensional (zones), one-dimensional (curves), or zero-dimensional (points). Each dimensionality requires different display forms, in accordance with accepted engineering practice.

Scalar two-dimensional fields are conventionally represented by drawing *contour plots* or *maps*. Contour maps are an elementary item in the visual vocabulary of the engineer, indeed most youngsters learn to read topographic contour maps long before they even hear of magnetism. In magnetics, such maps provide the ordinary, widely understood, flux plots. They are so commonly understood, and so commonly demanded, that some magnetics CAD systems produce them automatically unless specific instructions are given to the contrary.

A second widely used representation of scalar fields is the *zone plot*, in which zones of different prescribed characteristics are colored with different colors. The principle, once again, is familiar to everyone from topographic maps, where hypsometric tints are commonly employed to denote elevation. The advantage of zone plots is that they are useful for depicting discontinuous functions. Contour plots are readable only if the function being plotted is at most locally discontinuous, so that an implied interpolation (by eye) is possible. It is noteworthy that topographic maps of very rough, mountainous terrain often employ both hypsometric tints and contour lines. The loss of interpretation ability that results from discontinuities is in fact known as the *cliff problem* among mathematical cartographers.

Unfortunately, there is no conventionally accepted representation of a vector field. For example, the force on a spherical magnetizable particle placed in a magnetic field is readily enough calculated as

$$\mathbf{F} = K \text{ grad } (\mathbf{B} \cdot \mathbf{H}), \tag{14}$$

where K is a constant dependent on the characteristics of the particle. The magnitudes of the $\mathbf{B}$ and $\mathbf{H}$ fields can be illustrated, more or less comprehensibly, by drawing contour maps. But illustrating and communicating the force field graphically is much more difficult because there is no conventionally accepted representation for it—even though this force field is conservative, and it is not at all hard to invent reasonable graphic depictions of it.

Interactive and Batch Postprocessors

Two main computational approaches have developed for result evaluation, one essentially batch-oriented and the other interactive. The batch approach requires the user to specify the desired output before the solving phase occurs—the preprocessing is then done within the solver program and its output consists of the already postprocessed data. This approach is usually appropriate if the solution process is run on a remote, central computer. The interactive technique, which will be discussed here, allows the user to manipulate the output from the solver after completion of the solver phase. Its key ingredient is *second-thought capability*, a vital point because many questions to be asked about a field solution are prompted by the answers to earlier questions, or by the solution itself, and cannot be anticipated.

In many of the systems available at present the distinction of operating with different data types is hidden either in the command language presented to the user or in the batch approach that has been chosen. Thus, for example, in the MAGGY postprocessor the user specifies, before the solution is obtained, that a plot of FLUX is required, that the values of H along a contour should be produced, and so on. These commands are implemented by performing many of the operations described above but without the user being aware of the techniques used, the sequence of events, or the possible numerical difficulties. A similar situation obtains for the PE2D system.

Some of the more modern postprocessors provide the user an interactive, conversational form of communication. Examples of such systems are the MagNet postprocessor MagPost, and Ruthless—a general-purpose postprocessing system which allows the user complete control over the methods used in calculating the output data. This chapter is concerned with systems of the MagNet or Ruthless type since these are likely to become commonplace in the future and, essentially, form the basis for all the other postprocessing systems. The intention is to provide both an insight into the types of operations which may be performed and their associated data structures, as well as indicating the properties of the user interface and the graphics displays which are essential in the comprehension of the results.

Graphic Input

Graphic input is essential to interactive computing. In batch processes, on the other hand, it is redundant since batch processing generally neither requires nor permits user intervention while a program is running.

Although graphic input is widely used for a variety of purposes, it is really essential for only one kind of operation: the identification of structural elements within a model by pointing at a display. For example, to request information about the potential value at some spot, it is much more agreeable to issue a command which says, in effect, "please tell me the potential *here*", rather than to say "please tell me the potential *at the point x = 0.77524, y = −1.996*". In the former case, the user must of course point at the appropriate place, so that the word *here* is really understood to mean *the spot at which I am now pointing*. The actual physical pointing is done by using some device such as a light pen, stylus and tablet, trackball, or mouse.

The communication requirements of pointing are actually not quite so straightforward as might seem at first glance. Point identification can occur in any of three basic forms, not just one. When pointing at some spot, the intention may be to identify one of the follwoing:

1. The exact spot actually being pointed at.
2. An already defined point within the model and near the spot being pointed at.
3. A point exactly on a curve or line, and as near as possible to the spot actually being pointed at.

All three occur in practice, though in different contexts.

The first of the three cases rarely occurs when point values of numeric entities are required, for it does not often happen in practice that the flux density, or potential, or any other numeric value is wanted at a place defined no more precisely than is possible by freehand positioning. The second case, identification of predefined points, is very common. When inquiring after local data or recording data values, most designers prefer to specify precise locations definable in terms of the objects being dealt with, for example, "the upper left corner of the iron core". So far as program construction is concerned, this form of pointing is closely related to the first, since it is only necessary to find the coordinates actually being pointed at and to determine which of a set of predefined points is nearest. The third form is algorithmically a close relative of the first two as well.

On the other hand, the three appear in quite different contexts and different uses are made of them. From a user's point of view, they therefore appear as distinct entities.

It is interesting to note that in postprocessing, graphic input and graphic output serve radically different purposes. Graphic output invariably deals with data furnished by the solver program or programs—albeit usually in considerably reprocessed form. Graphic input very rarely, if ever, inserts new data; it generally serves to control the processing operations. Thus, graphic inputs are closely bound up with the control and command decoder programs that must exist in every interactive postprocessor, while the graphic output functions communicate principally with the calculators that do the actual data processing.

System Organization and Control

In view of the requirements outlined above, a postprocessing system for magnetic field problems should comprise three arithmetic units—one each for computing two-dimensional, one-dimensional, and point quantities. It must include displays appropriate to each of the three and must run under the overall control of a command decoder, which should preferably be conversational rather than batch-oriented. The system structure thus assumes the outline form of Fig. 1.

It should be noted carefully that the postprocessor communicates only with files emanating from the solver program, not with the program itself. If there is any information to be passed from one program to the other—a message about convergence failure, say, or an accuracy estimate—it must be passed by writing an appropriate message into the data file. As discussed in greater detail elsewhere in this book, restricting interprogram

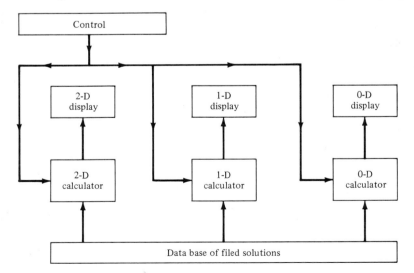

Figure 1. Basic postprocessor structure.

communication in this way achieves a high degree of program independence and permits alteration or improvement of one program without any effect on the other.

Communication between the data base and the calculators is shown in Fig. 1 to be unidirectional. This does not imply that the postprocessor cannot create temporary files and save intermediate results; but it does say that the postprocessor is not permitted to modify any data in the solution file. This restriction permits the same file of solutions to be postprocessed repeatedly, by using identical processes each time; it also makes possible the use of several different postprocessors in any desired sequence. Since none of the postprocessors is permitted to alter the data file in any way, none can tell whether the data file has already been postprocessed, and by which programs.

Mathematical Postprocessor Operations

A postprocessing system of the structure described above includes at least three mathematical calculation modules, which may be thought of as high-powered hand calculators capable of operations on fields. The manner in which these calculators operate will now be examined in somewhat greater detail than given above. Since the numeric calculator is the simplest to describe and has a reasonably limited set of operations it will be considered first. The general principles outlined, however, apply to all three calculators, as will become quickly evident.

The Numeric Stack Calculator

The calculator required to perform the operations described above can take one of two forms. These correspond quite closely to two of the more common computer structures encountered in hand calculators. The first form is a *register-based* machine in which every item is given an identifying name and arithmetic operations are performed by referring to the name. This is the type of calculator presented to the user by most high-level computer languages and is also one that is physically implemented in many of the more modern computer designs. Thus, to add two numbers the user must first store each of the numbers in named locations by some form of *assign* or *move* operation:

$$A = 1 \quad \text{means:} \quad \text{move the number 1 into location } A$$

and

$$B = 2 \quad \text{means:} \quad \text{move the number 2 into location } B$$

Addition is then performed by adding the contents of the two locations and placing the result in a third location:

$$C = A + B \quad \text{means:} \quad \text{add the content of } A \text{ to that of } B$$
$$\text{and place the result in } C.$$

The result in C may be displayed by writing the contents of location C to the current output device. In hand calculators there is generally one output device, the numeric display, and C will be shown in this display automatically. If there are several output devices, some additional commands may be required to select the right one.

The register-based machine has both advantages and disadvantages. In its favor is the fact that, once stored, the data in A and B remain until the user deliberately changes them. Unfortunately, the user has to keep track of what variable names have been used and what is in each location, so that he can perform some sort of clean-up whenever he requires extra memory for a new calculation. This type of machine may be very restricted by the available memory and cumbersome in terms of the language needed to operate it.

An alternative calculator structure is known as a *stack-configured* or *stack-structured* machine and is the form implemented in several popular hand calculators. The central idea in this case is that of a stack of data in which the currently active memory location is the top of the stack and the stack is as deep as required, or as the available memory will permit. Thus, entering a data item places it on top of the stack and data can only be removed from the top of the stack. The operation described above, that of adding 1 to 2, would be performed by the following sequence of operations:

Enter 1 places the number 1 on top of the stack:

current top $\longrightarrow$ $\boxed{\quad 1 \quad}$

Enter 2 places the number 2 on top of the stack, pushing the 1 down and making the stack correspondingly deeper:

current top $\longrightarrow$ $\boxed{\begin{array}{c} 2 \\ 1 \end{array}}$

The *add* operation now requires no operands because implicitly they are assumed to be the top items on the stack. The effect is to remove both the 1 and the 2 from the stack to an accumulator, add them, and replace them with the result on top of the stack. Since two items (addend and augend) are removed from the stack but only one (the sum) is pushed onto the stack, the depth of the stack shrinks by one:

current top $\longrightarrow$ $\boxed{\quad 3 \quad}$

Note that both operands of the addition operation have been deleted in this process. Thus, the user of a stack-configured calculator has to save copies of data items deliberately if they are not to be destroyed in the computation. However, the major advantage of this structure is the simplicity of the language used to operate it. The user no longer has to specify the operands of any operation, they are implicitly the top and next items on the stack for a binary operation and the result is always on top of the stack. The result of any computation may be displayed automatically because it is always in the same place. In addition, memory space is conserved because most operations do not increase the size of the stack. Any information previously in the stack is unaffected, for it remains resident at some lower level of the stack throughout.

For the CAD user, the simplifications in the command language and the automatic clean-up of unwanted data items make the stack-configured calculator extremely attractive. The operations on such a stack may be classified as unary, that is, operating only on the top item on the stack, or binary, that is, operating on the top two items on the stack and reducing the stack depth by one item. For the remainder of this discussion it is assumed that the type of calculator used in an actual postprocessor system is stack configured.

The numeric stack is the simplest stack in any postprocessing system—it handles single numbers and provides those operations which may be found on most hand calculators, i.e., the four operations of scalar arithmetic together with trigonometric and logarithmic functions. In addition, since the user need no longer specify locations in which data should be stored, it provides a set of stack management operations. These include entry of data onto the stack, deletion of data from the stack, duplication of the top item, and rotation of the stack. In addition to displaying the top item on the stack, it is often useful to provide a display of the current status of the data and to allow the user to name each item in order to keep track of it. This latter feature is an extremely difficult one to apply in general since it is not easy to determine an identifier for the result of an implicit operation such as an addition!

The Field Stack

The concepts discussed above, and detailed for the numeric stack, are not limited only to the case of single numbers. There is no reason why the two-dimensional field data structures output from the solver should not also be handled by a stack-configured calculator. Each item on the stack becomes an array representing a two-dimensional field, and any operation applies to the entire field currently on top of the stack.

When using the field stack, the designer need not be aware of the fact that fields are stored as arrays. To the user, solutions are known by name and referred to by name; it is the system's task, not the user's, to keep track of which names refer to arrays, indeed to which kind of arrays if there are several. In fact, each item on the field stack may have the properties of a multi-component vector field so that it becomes possible to handle three component fields, as well as scalars. A typical example of how this system might be used to calculate the x component of flux density is given below. It should be noted that any items previously on the stack (denoted below by *old*) remain undisturbed although the stack above them grows and shrinks.

Push on a single component vector potential field $1_z A_z$:

current top $\longrightarrow$
$1_z A_z$
old

Take its curl

current top $\longrightarrow$
curl $1_z A_z$
old

The result is a two-component vector field, equivalent in storage to two scalar arrays:

current top $\longrightarrow$
$1_x B_x + 1_y B_y$
old

Push on a unit vector in the x direction

current top $\longrightarrow$
$1_x + 1_y 0 + 1_z 0$
$1_x B_x + 1_y B_y$
old

and form the scalar (dot) product:

current top $\longrightarrow$
B_x
old

The top item on the stack has now been replaced with the x component of flux density. It may in its turn be used in further calculations, possibly involving data already on the stack.

As with the numeric stack, the field stack requires binary, unary, and stack management operations to be of use. The stack management functions are in principle similar, while the mathematical operations are considerably more extensive. In particular, it should be noted that in the simple example given above, the stack contains from time to time mixtures of scalar and vector fields, which require different amounts of storage and different organization.

The Curve Stack

The curve, or contour, stack is a special intermediate case between the field and numeric stacks. It is intended to handle data extracted from the field stack along a predefined contour and as such the basic data structure consists of one-dimensional arrays. Its operation is very similar to those described for the field and numeric stacks, having a command set which is smaller than the former but considerably larger than the latter.

Since the data on the stack relate to a one-dimensional geometric entity they can be only scalar in nature. Often directional differentiation of the field is required along a defined contour. This operation really belongs to the field stack since two-dimensional information is required for any direction other than along the contour. In practice, the operation may be performed during the transfer of data between the field and contour stacks.

Implementation Examples

At least two general-purpose electromagnetic field postprocessors of the type described are in reasonably widespread use. These are the postprocessor for the MagNet system (MagPost) and the Ruthless postprocessor. These are both examples of the type of system described in this chapter. The initial target hardware systems and the types of solution system are somewhat different in the two cases. Ruthless was intended to interface to very large, high-order finite element programs. It provides the user with the basic vector operations and relatively sophisticated graphics displays. The system runs on a Prime or VAX-11 computer and thus has access to very large virtual memory spaces. MagNet was implemented initially on a microcomputer system and thus was written for an environment with limited memory resources. More recently, it has become available on a wide range of computers both large and small. Its intended user community was industrial designers and, consequently, considerably more emphasis was placed on the programmability of the system so that the basic operation of the system could be hidden from the average user while still providing a sophisticated set of operations for the research engineer.

Every finite element analysis system has some form of postprocessing subsystem associated with it, although many present the user with a pre-cut set of operations which do not allow the flexibility of calculation or display which has been described in this chapter. Typically, the user of such a system might be expected to specify the kinds of graphs desired before the problem is solved. Examples of this form of system are furnished by MAGGY, AOS-Magnetic, and one of the PE2D processors. These postprocessors are not interactive, but should be regarded as interactively launched batch programs. Others provide complete generality by producing a printout containing all the numbers produced and leave their manipulation up to the user. However, such systems are becoming rarer and newer designs, which adhere to some of the interactive concepts but exhibit a strong batch background, are becoming more commonplace.

A Postprocessing Language

The MagPost subsystem of MagNet presents the designer with a system of the type described above and provides an example of a language designed with three stack-configured calculators in mind. As with other parts of the MagNet system the user is prompted for input by a message

```
>> FIEL ready.  Command:  _
```

However, in contrast to other parts of the MagNet system, FIEL is not the currently active model or problem. Instead, this first word in the command prompt indicates to the user which calculator is currently operative and thus, implicitly, which set of commands is currently available. On startup the system always makes the FIELd stack active.

Several commands are provided for stack management and switching between calculators. They may be used at any time and apply to all the calculators:

STAC*k* **name**	causes the **name**d calculator to become active.
DUPL*icate*	creates a second copy of the top field.
DISC*ard*	removes the top field.
EXCH*ange*	exchanges the top two fields.
ROTA*te*	causes the stack to rotate.

The data base provided as input to MagPost contains the geometric data, the potential solutions, and various constants defined by the user at the MagProb stage. These data sets are therefore only applicable to the field or numeric stacks. Before work on a solution can start, that solution must be placed on top of the field stack. Thus a DIRE*ctory* command which lists the current contents of the MagNet data base on the terminal is required and, having chosen the appropriate solution, it may be placed onto the stack by means of the command GET. For example, if the required problem was named TRSC then the command sequence would be

```
>> FIEL ready.  Command:   GET TRSC
```

As with the hand calculator, it is important for the user to be able to determine exactly what data is available to work with. Whereas on the numeric stack or in a hand calculator only a single number is needed to provide all the required information, the field calculator presents a problem. In general, the user does not want to know the value of the top item at every point in space; removing the need to dump all values to the printer is the main aim of a postprocessor! However, it is still necessary to have some idea of what is on top: Is it vector or scalar? If a vector, how many components? Thus, what is really required is a *status* display of the data on the stack; useful information might include an indication of the problem name, the order of polynomial approximation currently used, the geometry that the solution relates to, and the approximate size of the field in terms of the amount of memory used. These data may vary from system to system but are basically the ones required to allow a user to keep track of the field being manipulated.

A typical status display might be invoked by the command STAT*us* and might display something similar to the following:

```
RX IX RY IY RZ IZ VECT GEOM ZONE SOLU  COPY PERCENT DISK
                  SCAL NAME NAME NAME        SPACE USED
Z  Z  Z  Z  V  Z  VECT LEAK *NUL TRSC   1      10
```

Although it may seem a little confusing at first glance, this status display informs the user that the solution to a problem called TRSC defined on a geometry LEAK is on top of the field stack and that this solution consists of a *z* directed real vector field. This field can be DUPL*icated* to place a second copy on the stack. Afterward, the status display will contain

```
RX IX RY IY RZ IZ VECT GEOM ZONE SOLU  COPY PERCENT DISK
                  SCAL NAME NAME NAME        SPACE USED
Z  Z  Z  Z  V  Z  VECT LEAK *NUL TRSC   1      10
Z  Z  Z  Z  V  Z  VECT LEAK *NUL TRSC   2      10
```

Note that the copy count has increased, indicating that both fields are the same. If the CURL operation is now used, the top field on the stack is replaced by its curl:

```
>> FIEL ready.  Command:  CURL
>> FIEL ready.  Command:  STAT
```

producing as the status display

```
RX IX RY IY RZ IZ VECT GEOM ZONE SOLU  COPY PERCENT DISK
                  SCAL NAME NAME NAME        SPACE USED
V  Z  V  Z  Z  Z  VECT LEAK *NUL         1      7
Z  Z  Z  Z  V  Z  VECT LEAK *NUL TRSC    1      10
```

Clearly, the DUPL and CURL commands have now produced a two-component vector field on top of the stack which represents the curl of the original. If the original field was a vector potential then the new item consists of the flux density. Because the new field is not the same as the original the solution name must have altered but, since there is no information about a new name, the system sets the name to four blank characters. One interesting point is that the amount of disk space used has decreased even though the resultant field has more components than the original. This curious fact stems from the precise fashion in which data are represented internally, a matter discussed in some detail later.

The lack of a solution name may confuse the user, but then so might the original name if it had been employed to identify the new item! Unfortunately, there is no easy way of deriving a new name based on the original name and the operation; how can the result of FRED added to TRSC be described? To reduce chances of confusion, the MagPost language allows the user to specify the name of the resulting field. Thus the above operation might have been executed by

```
>> FIEL ready.  Command:  CURL BFLD
>> FIEL ready.  Command:  STAT
```

giving

RX	IX	RY	IY	RZ	IZ	VECT SCAL	GEOM NAME	ZONE NAME	SOLU NAME	COPY	PERCENT DISK SPACE USED
V	Z	V	Z	Z	Z	VECT	LEAK	*NUL	BFLD	1	7
Z	Z	Z	Z	V	Z	VECT	LEAK	*NUL	TRSC	1	10

The language of MagPost adheres to the same rules of structure and syntax as the other parts of the MagNet system. Commands are, in general, those which naturally describe the operations defined earlier. Thus ADD adds the top two fields on the stack, and the operation is consistently called ADD over all the stacks.

Interstack communication is provided by a scratch area accessible to the top of each stack and to the various geometric data bases accessed by the system. The scratch pad provides a buffer for storing data during a change in calculator. For example, a user can define a contour within the two-dimensional data base by using the CONTour command. The contour is saved in a data base of one-dimensional geometric entities, with the specified name, in this case POLE. If no name had been given, the system would have used the system default name. One contour is allowed to be the system default contour at any time.

```
>> FIEL ready.  Command:  CONT POLE CREA
```

The values of the field currently on top of the field stack along the contour POLE can then be obtained by

```
>> FIEL ready.   Command:   EXTR POLE CONT
```

This command places the values in the scratch pad ready for pushing onto the curve stack. The calculators are then switched:

```
>> FIEL ready.   Command:   STAC CURV
>> CURV ready.   Command:
```

and the data are pushed onto the curve stack by

```
>> CURV ready.   Command:   ENTE MAGN
```

This sequence of commands will create a curve named MAGN on top of the curve stack. The transformer equivalent circuit calculation referred to in the previous chapter provides a typical sequence of operations, as will be seen below.

The Transformer Example

As described earlier, the leakage reactance of a transformer may be obtained from the short-circuit test simulation by calculating the total energy. This requires a surface integration over the entire model, as discussed elsewhere in this book. Using the MagPost language and calculator structure, the operations for this calculation, assuming that the short-circuit simulation results are stored in TRSC, are as follows.

```
>> FIEL ready.   Command:   GET TRSC
>> FIEL ready.   Command:   GET TRSC DENS
>> FIEL ready.   Command:   EMBE
>> FIEL ready.   Command:   DOT  AYJY
>> FIEL ready.   Command:   INTE
>> FIEL ready.   Command:   STAC NUME
```

The first two commands place TRSC on the field stack, followed by the current densities associated with it. EMBEd increases the order of approximation by one polynomial degree, so that the current densities which initially were of degree zero are now expressed as first-degree polynomials. The DOT command then forms the scalar product $\mathbf{A} \cdot \mathbf{J}$; INTEgrate computes the integral of this quantity and places its value on the numeric stack. At this point the numeric stack contains twice the total energy in the system. The calculation of inductance requires the input of the primary current. This could be obtained by direct entry onto the numeric stack, or else by asking for an integral of the input current over one of the windings. If the input current was 10 amperes, the final value of inductance can be obtained by the commands

```
>> NUME ready.  Command:  ENTE 10.0
>> NUME ready.  Command:  DUPL
>> NUME ready.  Command:  MULT
>> NUME ready.  Command:  DIVI
```

Here ENTE*r* creates a numeric entry of 10 amperes; DUPL*icate* and MULT*iply* compute its square. This quantity resides above the computed double energy value in the numeric stack, so that DIVI*ding* one by the other causes the top of the numeric stack to contain the desired inductance value. It may be displayed by using the STAT*us* command to show the current value on top of the NUME*ric* stack.

A similar sequence of operations applied to the open-circuit simulation will produce the combination of the leakage and magnetizing reactance. The core loss term in the equivalent circuit requires a set of operations which includes the evaluation of the flux density squared (as a lookup parameter pointing to the loss curve), and a request for the appropriate value for each element by accessing the required auxiliary material curve, e.g.

```
>> FIEL ready.  Command:  VALU TRSC AUXI
```

In all the above calculations, numeric output is the natural form of result presentation. On the other hand, many of the results of an analysis may be better displayed in graphical form, as discussed next.

Graphic Displays of Results

The very essence of interactive computing is to see what operations are done as they are done, and to see what results are produced, as they are produced. Any interactive postprocessor implementation must therefore make provision for extensive graphic displays and graphic inputs.

Displays, Stacks, and Formats

Each of the calculators described above deals with geometric data of a particular spatial dimension and in each case there exists a familiar, conventional data representation. Thus, the field stack has a natural form of display which is based on the two-dimensional drawing of the problem geometry, while the natural display format for the curve stack is the graph, in which field data are plotted against a variable coordinate extracted from the geometry. The numeric stack requires little more than a numeric display!

In an ideal world each calculator should therefore have its own display. This arrangement would require two graphic displays and one numeric display for the system, and would probably cause considerable confusion for the user! Instead, the displays may be considered to be logical devices which are present but not connected to the physical display device, the connection being made whenever they are actually needed. Thus, when the user is working with the field stack the display is assigned to the top of the field stack and shows the current top-of-stack item. Similarly, when the curve stack is the active one, the display shows the current state of that stack. This was, essentially, the approach used in postprocessing systems such as Ruthless. In MagPost only a single physical display device is used, but its drawing surface is thought to consist of a set of windows each of which may be user defined and attached to a calculator. In this arrangement a calculator may have several windows allowing several aspects of a problem to be displayed simultaneously. Each window is assigned to one item on the stack. The exact format and handling of the display may ultimately depend on the actual graphics device being used.

The Field Calculator Display

Although numerical results provide a designer with detailed information about the performance of a magnetic device, considerable quantities of qualitative data are available from a field plot for a two-dimensional analysis. The pictorial representation of the field can provide information about both the correctness of the data provided to the solution system and the field intensities in various parts of the device. For preliminary design work this representation of the field may well be sufficient to indicate possible changes to the design without any numeric data being needed. There are various different ways of displaying field data pictorially; contour plots (flux plots) are common but other representations may be useful also. Three different forms of display are shown in Fig. 2: a contour plot, a plot of magnitude against distance along a cut through the model, and a "hill-and-dale" quasi-three-dimensional plot.

The definition of the vector potential $\mathbf{A}$ in terms of the flux density $\mathbf{B}$ means that, for a translationally uniform two-dimensional geometry in which $\mathbf{A}$ has only one component, lines of constant A represent lines of magnetic flux. Conventionally, $\mathbf{A}$ is defined implicitly by

$$\text{curl } \mathbf{A} = \mathbf{B}. \tag{1}$$

For an elementary path or loop in which the flux density $\mathbf{B}$ is nearly uniform,

$$B = \frac{\phi}{lw}, \tag{15}$$

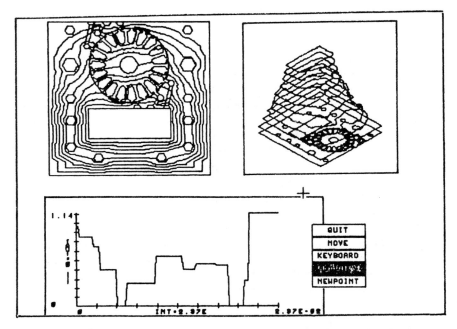

Figure 2. A typical multi-window display, showing a variety of graphical presentations.

where ϕ is the magnetic flux, and lw represents the cross-sectional area of an elementary loop. In a two-dimensional, Cartesian geometry, l may be taken as unity. Substituting from equation (15) into equation (1) gives

$$\phi = \omega A \delta. \tag{16}$$

If lines of constant vector potential are drawn so that the difference in A between adjacent lines is constant, they define flux paths through the device. The lines of constant vector potential may therefore be regarded as flux lines.

It should be remembered, as described in previous chapters, that $\mathbf{A}$ is a potential function; its absolute values have no particular significance. In fact, they are undefined unless the value of $\mathbf{A}$ has been specified at some point in the model prior to solution. This is a result of the infinite number of potentials that can satisfy equation (1), each differing by the gradient of a scalar. However, local differences in $\mathbf{A}$ are physically significant. Thus, although a flux plot (or actually an equipotential plot) provides qualitative information, the absolute potential value of each line may mean little.

The above argument holds for two-dimensional problems solved in the x-y plane, but it does not apply to axisymmetric problems. In the latter case equation (1) is modified in an axisymmetric coordinate system to become

$$\mathbf{B} = \frac{1}{r} \left[\frac{\partial}{\partial z} (rA) \, \mathbf{1}_r - \frac{\partial}{\partial r} (rA) \, \mathbf{1}_z \right]. \tag{17}$$

Comparing with the Cartesian version of the same equation it can be seen that if the potential function A is replaced with rA in an axisymmetric system then the equipotentials once again describe flux paths.

To obtain plots of flux distributions in two-dimensional magnetics it suffices to draw contours of constant vector potential A in x-y problems, or of the product rA in r-z problems. In practically all finite element based programs, the equipotentials are drawn on an element-by-element basis, not by following lines corresponding to specified potentials. The procedure of plotting on an element-by-element basis of course produces the same end result as following flux lines around. However, it produces plots by drawing bits of lines in unconnected locations, and it can be a little startling when first encountered in interactive systems, where the plot is formed while the user watches.

The postprocessing phase of the design system has access to both the geometric data base, describing the element configurations, and the potential solution. In addition, the approximating polynomials used over each element are known. Using these data the potential may be calculated at any point within an element. More significantly, for plotting equipotentials, the intersection point between a specified potential value and an element edge can be quickly determined. Thus, if the edge intersection points are found, and the variation within the element is known, the equipotential can be plotted. Typical flux plots have been used as illustrations throughout this book. In all such plots, regions of high flux density are of course visible as regions of crowded lines.

Alternative Field Representations

Flux plots are not the only form of display which may be required for the field stack. In describing and displaying flux plots it has been implicitly assumed that the geometry can be displayed and, in asking questions of the solution, this may well be an extremely important display. In addition to the outline, the element distribution can provide information as to whether the flux plot is showing features of the solution or artifacts introduced by a bad discretization.

Another form of display has already been mentioned, the intensity plot. For many calculated parameters, such as loss, energy distribution, and maybe even the flux distribution, an intensity plot can provide considerable information. This form of display is common in systems for mechanical and civil engineering and is becoming increasingly common in electrical engineering. An example of such a plot is shown in Fig. 3.

Many magnetic device engineers are accustomed to have another set of lines drawn, orthogonal to the flux lines, generally called *scalar equipotential lines*. Historically, this habit probably dates from the technique of manual flux plotting, by which many problems were solved in the 1920s and 1930s. Scalar equipotential lines are not derivable from a vector potential solution as easily as the flux lines; indeed, in current-carrying regions they cannot be drawn at all because the magnetic scalar potential is undefined there.

Figure 3. A region plot of flux distribution. Gray shades often communicate better than assorted colors.

In axisymmetric problems, where flux lines and vector equipotential lines do not coincide, the designer has a choice as to which to plot. Most classically trained people probably prefer flux-line plots, but in the past decade or so various examples of equipotential plots have appeared in the literature as well. Indeed, this book contains several examples of axisymmetric vector equipotential plots.

The Curve Stack Display

The flux plotting description discussed the value of pictorial representations of the field potentials. This form of display can provide information in an easily understood manner because it is a conventional style for the data. Similarly, there are conventions for displaying other forms of data which make it easier to absorb the information being presented.

Several sections of the preceding chapter stressed the need for defining contours within the geometric data base and for extracting values along them. Indeed, a calculator has been described for manipulating data in this form. The conventional method of presenting data calculated as a function of one variable is by means of a graph. When presented in this way salient features can be spotted quickly; if the results are produced as a set of numbers such features are not as easy to find.

A graph implies that two related variables are to be displayed, one against the other. Frequently, in an electromagnetic device this may be the distribution of some parameter along a defined contour, or against one of the coordinate values of the points on the contour. However, this may not always be the case and it may be possible that the user needs to plot two data items against each other, neither of which is the contour itself. This might occur if an impedance diagram is being constructed. Thus, the general form of the graphing operation plots the top two items on the curve stack against each other.

Before a graph can be plotted, several parameters must be specified such as where on the graphics display the graph should appear, the axis types to be used (logarithmic or linear), the axis labels (if any), and so on. Each of these should be alterable by the user although a default set of values is required to bypass the set-up operation.

A second format for data along a curve might well be tabular. This form of presentation is essentially numeric in nature and engineers generally find it more difficult to understand than graphs. A typical curve, for example, might have 500 values along its length. However, if further processing of the data is required, the tabular form is often preferable because it is readily usable by a computer.

Cosmetic Improvement of Graphs

In all graphic output operations, two quality evaluation criteria oppose each other. Graphs as drawn by a postprocessor are illustrations of quantities found in an analysis, and it can be argued that they ought to be exact representations of the results produced by the calculations that constitute the analysis. On the other hand, graphs are there to communicate physical information to the user, so it can equally well be argued that any approximations or computational artifacts that arose in the analysis are defects in the solution and ought to be removed from the plots. In other words, the first viewpoint asks for an exact plot of the approximate solution—no matter how obvious the errors of approximation; the second one suggests elimination of at least the most obvious physically impossible kinks and jags in the solution, through some well chosen process of smoothing.

Whether graphs should be smoothed or not is a tricky question. Engineering analysts generally prefer that no smoothing be applied, on the grounds that visible defects in a solution are highly informative about how far the solution should be trusted. On the other hand, kinky flux lines and jagged graphs do not inspire confidence, particularly when presented at management meetings or customer presentations! This problem can be particularly nasty when presenting flux plots obtained from finite element analyses, for these almost invariably show very smooth, clean contours in regions where local energy density and flux density are high, but become ragged and messy where flux lines are far apart. After all, the very essence of finite element analysis is a minimization of *global* error, so high precision is to be expected where flux density is high, and vice versa. Unfortunately, areas of low importance often fill a large proportion of the picture space in plots.

The purpose to be served by graphic plots might be the best guide to choosing whether any cosmetic improvement, by smoothing or otherwise, should be made. Before making the choice, however, it should be observed that good smoothing is surprisingly difficult. The smoothing algorithm must not introduce any visually offensive features, especially if these might be in conflict with expected or known physical principles. For example, a smoothed flux plot of a harmonic field must not show any closed lines that form "islands" where there are no sources. Similarly, smoothing that removes kinks in flux lines must maintain kinks at material interfaces. Violations of physical expectations in this way are equally readily spotted at a customer presentation, but much more likely to destroy confidence than the unretouched graph.

The Numeric Stack Display

The final form of data presentation is numeric, a natural form for the computer but not always as useful for the engineer. In general, numbers of use relate to global quantities such as the terminal parameters discussed earlier, or to single values such as the flux density at a specific point in the device. The numeric data are important to the designer if presented in these limited forms because they provide selective detail of the field solution.

For all these forms of output, archiving systems are required to save data for further processing or comparison with other designs.

Graphic Input

As with other parts of a CAD system, such as mesh generation and problem definition, the user can communicate better with the postprocessor by being able to query the system graphically. This requires a certain amount of graphic input capability. For example, if the value of the field at a specific point is desired the user could identify the position by entering the point coordinates at the keyboard. This procedure is both tedious and time consuming if the intention is to get a rough feeling for the solution by looking at several points. A better approach is to let the point be specified graphically by "pointing" at the desired location within a display of the geometric model on the graphics screen. Of course, the numeric approach is still essential if very accurate positioning is needed, so a good postprocessing system must provide both methods.

Once the user has entered into a graphic pointing mode, for example to define a graphic entity, user communication is continued via graphical menu systems. The user thus only employs one input device at a time (graphic or keyboard), and only one output screen (or screen window) at a time. Confusion is eliminated by providing structural simplicity.

Points and Contours

The specification of a point is simplicity itself: the graphic cursor is steered to the appropriate place on the screen, using the graphic input device, and a button or key attached to that device is pressed to signal that the intended point is at the present cursor location. If the input device is a stylus, for example, the usual method is to activate a pressure-sensitive switch in its tip by pressing down as if the stylus were a pencil. Pucks, mice, and other movable devices usually contain one or more pushbuttons or switches for the same purpose.

Signalling the cursor position is not enough to identify a point. Depending on the context, the cursor location may be interpreted as being the exact point intended; such might, for example, be the user's intention when asking for the flux density at a point. However, in other situations the user may wish to point to a previously defined place, such as the corner or some structural feature. Distinguishing one intention from the other can be achieved through context dependence or, preferably, through menu selection.

Contours, like points, can be indicated by pointing at the graphics screen. A contour composed of straight-line segments can be specified easily, as a sequence of points implicitly understood to be connected by lines. If a contour is required to fit a complex shape, such as around the tooth of an electric machine, shapes more complex than straight lines may be needed. Complicated curves of this sort can be built from a richer set of primitive segments, including at least circular arcs. As with points, the exact location of a contour may be important and some of the fiducial points on the contour should be entered via the keyboard or attached to defined nodes. Since a point is both a subset of the two-dimensional and the one-dimensional data bases it is also reasonable to allow point definitions on graphs of data defined over contours.

Zone Definition

Frequently, to analyze a design the user has had to solve not just for the field in the section of interest in the problem, but also a considerable part of the surrounding geometry in order to provide reasonable boundary conditions. An example of this might be the solution of a pole pitch of an electric machine when the designer really requires only the air-gap solution.

One approach is to allow the specification of a section of the two-dimensional geometry as a new two-dimensional piece. The effect, from a user point of view, is similar to that of the graphical zoom in the mesh generator. However, the creation of a zone provides more than just the ability to perform geometric zooming. It allows the user to create a new problem on the field stack containing only those elements which relate to the piece of the problem in which he is interested. The advantage of the

EXTR*acted* field is that it is smaller than the original and thus processing operations run considerably faster. Hence the system may become more interactive. The ZONE NAME in the STAT*us* displays shown earlier in this chapter relates to this particular operation.

As with contours and points, zones may best be specified graphically by pointing at the parent two- dimensional display. In many situations the zone may replace the need for a zoom operation, however, the difference lies in the fact that one is to be used for calculation and thus must be a complete problem in its own right, whereas the other is purely a display feature.

The definition of a zone and the extraction of data related to it create a new field on top of the field stack. However, the definition of a contour or a point creates a data structure which cannot be stored on the field stack, because the geometric dimensionality of a contour or point is not the same as that of a field. When this happens data must move from one calculator to another. The mechanism for performing this type of operation was described earlier. Thus the graphic input system effectively provides a method for intercalculator linkage.

The Custom Postprocessor

Although the range of operations desired in a postprocessor is extremely wide, for any specific design project it is likely that either a small subset or a specific sequence of operations will be used frequently. Consider the transformer example: the operations required are complex but the designer may want to check the equivalent circuit for several designs. An ideal postprocessor would therefore have a set of commands which could be of use to a transformer designer. Another user might prefer a postprocessor which has commands relevant to electric machine designs of a particular type, and yet another might be interested in the characteristic impedances of transmission lines. It is possible to construct a processor for each of these requirements but it is likely that there will be as many different processors required as there are users. In fact, there may be more because the user's requirements may change from day to day! To provide this form of custom processor, without building a dedicated system for each user, the concept of a command shell is introduced.

The Command Shell

The shell in a postprocessor provides the user with a language for writing new commands for the system. Each basic command to the processor becomes programmable so that sequences of the basic commands may be built up. Once a sequence has been defined it may be stored with a user defined name and the name becomes a new command to the system. This is similar to the programming capabilities described for other parts of a CAD system.

As an example, consider the inductance calculation for the transformer based on flux linkages. Each step may be made into a part of a stored program by prefixing it with a line number. When a line number is encountered, the shell, that is, the command interpreter, stores the command in a program buffer rather than executing it. For example,

```
>> FIEL ready.  Command:  100 GET TRSC
>> FIEL ready.  Command:  110 GET TRSC DENS
>> FIEL ready.  Command:  120 EMBE
>> FIEL ready.  Command:  130 DOT
>> FIEL ready.  Command:  140 INTE
>> FIEL ready.  Command:  150 STAC NUME
>> FIEL ready.  Command:  160 PUSH
>> FIEL ready.  Command:  170 ENTE 10.0
>> FIEL ready.  Command:  180 PUSH
>> FIEL ready.  Command:  190 MULT
>> FIEL ready.  Command:  200 DIVI
>> FIEL ready.  Command:  210 STAT
```

The commands entered above are treated as though they are a program for the command decoder and are not executed as they are entered (note that the current stack does not alter even when the numeric stack is specified) but are stored in a command buffer, the contents of which may be edited by typing in commands prefixed with line numbers. The interpreter always executes the commands in ascending sequence of their line numbers. The sequence of commands may be stored in a library of defined verbs by using the VERB command:

```
>> FIEL ready.  Command:   VERB INDU
```

The VERB command converts the numbered lines that make up the stored program into a new command INDU. This command may now be used just as any of the other commands. It may, itself, be used as a part of a stored program allowing the construction of extremely complex commands.

The command INDU as defined above is somewhat restricted in terms of its input and output data. Although it performs all the correct operations for finding inductance it assumes that the input field is always in TRSC and that the input current is always 10.0 amperes. These restrictions are removed by provision of *symbolic place holders*, which appear in stored program lines in the form of numbers in brackets. In the above example, place holders (dummy parameters) may be used in place of fixedly defined field names or currents, as in the following:

```
>> FIEL ready.  Command:  100 GET (1)
>> FIEL ready.  Command:  110 GET (1) DENS
```

```
>> FIEL ready.    Command:   120 EMBE
>> FIEL ready.    Command:   130 DOT
>> FIEL ready.    Command:   140 INTE
>> FIEL ready.    Command:   150 STAC NUME
>> FIEL ready.    Command:   160 PUSH
>> FIEL ready.    Command:   170 ENTE (2)
>> FIEL ready.    Command:   180 PUSH
>> FIEL ready.    Command:   190 MULT
>> FIEL ready.    Command:   200 DIVI
>> FIEL ready.    Command:   210 STAT
```

If the above is converted into a new command INDU through use of the VERB command (obliterating any previously existing command by the same name), greater flexibility is gained. The new command INDU may now be given adverbs in the same way as any other command and the inductance can now be found by

```
>> FIEL ready.    Command:    INDU TRSC 10.0
```

In the execution of this command, the given adverbs are substituted for the symbolic place holders before the individual command lines are acted upon. The first adverb in the string takes the place of the symbolic place holder (1); that is, the string of four characters TRSC is substituted for the three characters (1). Similarly, the four characters 10.0 replace the string (2). It should be noted that the substitution is strictly one of replacing one character string with another; no attempt is made to check for validity until the time has come to actually execute an elementary command line.

Using the programming feature a user may construct a custom postprocessor tailored to his own particular needs which, eventually, may isolate him from any need to manipulate data at the vector analysis level. In fact, the operations of the batch processors may be implemented and the ultimate user of the system may be almost unaware of the software structures underlying the commands!

Heuristics and Graphic Inputs

Frequently, being able to program new commands in advance is only a partial solution to a user's problems. In the analysis of several similar designs it may be desirable to perform the same operations on each one but the exact sequence may not be known in advance.

Typically, a designer may require data at various points through a design, displayed graphs of values along specified contours, and flux plots over various zones. If all the designs have been built over the same geometric data base by merely altering the material definitions and the source densities, then defined contours and points will apply to all of them. In this situation it is often more useful to be able to "teach" the postprocessor exactly what is required in terms of output results.

The *heuristic* mode of programming may be used to persuade the system to "learn", i.e., to remember the sequence of operations required. Not only are keyboard commands stored in the command file in this case, but also graphic input. Since all graphic input tokens in MagPost are cursor hits, they must also be stored in that form. The program, acquired heuristically, may of course be stored just like any other command verb and executed in the same way.

Using the learning facility is akin to programming in an interpretive language while at the same time having the facilities of a stored program. The learnt verb can then be programmed to execute over an entire data base of solutions and the postprocessor can be left to operate in a batch mode in which the desired results are filed away on disk for later access by the designer. One difference between verbs created in the heuristic or "learning" mode of a system and those created by means of noninterpretive editing is that it is difficult to edit a learnt verb because of the cursor hit storage. Thus, any errors which occur whilst the system is taught the sequence of commands will be faithfully reproduced whenever the verb is executed!

CAD Systems Present and Future

Engineering is said to be the art of coordinating men, machines, and materials to produce useful goods and services. It includes both scientific and managerial components so engineering computing must reflect the characteristics of both. Development of the CAD art, from computer-aided analysis (which it was in the early 1970s) to computer-aided engineering (which it might become in the late 1980s), is moving it steadily closer to the professional practice of engineering. Software for computer-aided design, which in the seventies amounted to straightforward scientific applications programs, has increasingly begun to take on the system orientation associated with business operating systems in the past.

System Structure

Computer-aided design systems of the present have been shaped mainly by the traditions of scientific computing and by the constraints imposed by computing hardware. These influences have had a strong effect on working patterns of system users.

The following overview of system structures will attempt to review the forms taken by past CAD systems in magnetics and related areas, keeping in mind both historical practices and technical limitations. Some probable future directions will be suggested, as illustrated by current systems.

Files and Processes

Scientific computation frequently involves very complicated mathematical procedures but rather small quantities of input and output data. In contrast, commercial data processing tends to involve large numbers of extensive data files, with only a little processing work done on each unit of data. As a result, scientific computation has tended to focus strongly on the processes of numerical computation and has been *process oriented*, while commercial data processing has had a much stronger *file orientation*.

Processes and files thus occupy almost opposite roles in the two traditional forms of batch computing. So do attitudes of both program users and program builders: scientific computing tends to stress portability and universality of *programs*, while management computing people are much more preoccupied with transportability of *data*.

CAD in the Batch Era

In the traditional batch-processing era of computing, scientific application programs were generally regarded as single processes. They possessed one input stream and one output stream; users took care to load the input stream with correct data and hoped to receive the correct answers after program execution. The input data were usually numeric, and the output typically comprised from one to several hundred pages of numbers. All but a few of these served purposes of program verification, the remaining few being "answers" in the strict sense. The internal structure of the program was likely to be intricate, and program alterations were avoided if at all possible. If a new application required either changing the program or reformatting data, data rewriting would normally be preferred.

Business computing of the same period concentrated on creating files in well defined formats and maintaining archival security over the data. Programs to achieve specific data processing objectives were then created. When changes in taxation, commercial law, or labor contract made it necessary either to alter data file structure or to change a program, reprogramming would normally be preferred.

Because it is almost impossible to assemble the input data stream for a large task without errors, work in the batch-processing era was traditionally subdivided into a linear sequence of tasks or phases. Each phase was, or at least could be, separately executed and the results were often separately checked for correctness. Thus one spoke of geometric modelling as a separate task, the assembly of finite element equations as another task, and of their solution as still another. Within each task, subtasks naturally followed: geometric meshing or modelling quite naturally subdivided into a node specification subtask, an element specification subtask, a boundary condition subtask, and so on, with opportunity for separate checking of each set of intermediate results. This approach usually leads to a linear software structure with many intermediate files, as illustrated in Fig. 1. The intermediate files are all of an external character, i.e., man-readable, and so organized as to make sense to a human reader. The associated working pattern for the user is linear: he must run each process to completion, verify that the output data are exactly what he requires, then proceed to the next. Linear working patterns have the strong advantage that individual programs can be kept simple, therefore relatively easy to check out and debug. Furthermore, special-purpose programs can easily be inserted into the stream when the occasion demands. The data files can be made compact and few need be allocated storage space at a given

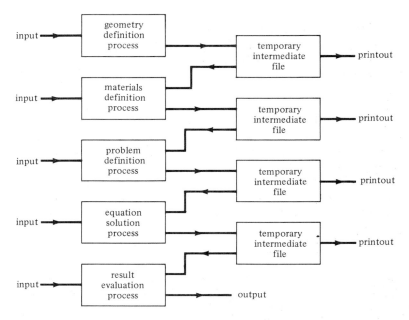

Figure 1. Process-oriented software structure. Each process follows sequentially on the last, using data provided by previous processes. Intermediate data may be introduced, but each process must be run separately.

time, since intermediate data sets can be discarded as soon as work moves on to the next process phase.

Batch computing with overloaded machines and slow turnaround times has invariably led to a desire to have as much computing done as possible between successive user accesses to the machine. The linear working pattern with frequent access to intermediate files, as in Fig. 1, has therefore always been popular but has never been considered particularly desirable. Instead, most system designers of the 1970s strove to delete intermediate files automatically, and to have the processes follow one another without user intervention. Some downstream processes might of course require additional input data; if so, the user could be expected to prepare it beforehand, but to have it read from a card reader or other sequential medium when required. The system structure thus took on an appearance such as that shown in Fig. 2.

Interactive Use of Central Machines

During the 1970s, interactive computing gradually became a reality. Typically, engineers had access to printing or video terminals connected to a large remote computer. Existing batch-processing-oriented CAD software was often pressed into interactive service by "just adding interactive front and back ends". These "ends" often amounted to simulating card reader

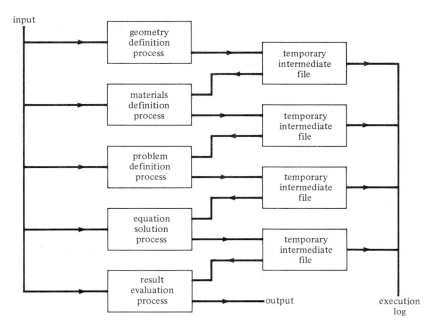

Figure 2. Mature batch CAD system. All input data are loaded into a single sequential stream and all required output issues in a single stream. Processes interlock, since many subprograms can be made common.

and line printer communication while providing some degree of error tolerance in such matters as numeric input formatting. As terminals with graphic display capabilities made their appearance, they were treated in somewhat the same fashion: they were made to simulate pen-and-ink plotters, batch-processing hardware to which engineers had already become accustomed. In other words, the linear working patterns associated with the mature batch-run program remained intact, even though some flexibility was gained in the individual phases of work. For example, the specification of graphic output formats ("how many contour lines do you want plotted?") was no longer always required at the beginning of a run, attached to the input data. Instead, such specifications could be delayed until the user had gained some idea of the intermediate results.

From the user's point of view, linear program structures are inflexible because they do not possess *second-thought capability*. That is to say, the user is unable to return to a previous phase once he has entered the next; he is required to traverse the various process steps in an inexorable downstream sequence. For example, in the structure of Fig. 2, he has no opportunity to change his mind or to correct an error in the geometry description data once he has proceeded to materials definition or beyond. Not only will the process sequence not allow going back; worse, the files in which the data now reside have a totally different structure from his original input, and he could not recognize his original data even if the process sequence were to permit him to access the files. Lack of second-thought capability is not a matter of concern in batch processing, since there is in any case no time or occasion on which the user could interfere with the computer run once it is launched.

Provision of second-thought capability is achieved by making file structures invariant under processing. That is not to say processes ought to leave file contents unchanged, of course, but only that the *structure* of a file should be identically the same before and after processing. To choose a simple example, suppose some process is to effect a renumbering of all points in a finite element model so as to accelerate later processing (e.g., to reduce time and memory requirements in equation solving by bandwidth reduction). In classical batch computing, the input (unsorted) file would likely have had a format and structure chosen to suit the program that generated the file, and an output structure convenient for reading by the next program. In an interactive system, it would be preferable to choose a single file format for all operations that deal with points. In this way, the user could still correct an erroneous coordinate value, even in the renumbered file.

Design of interactive CAD systems thus requires, first and foremost, design of suitable file structures. Processes are then set up so that file structures are maintained intact even while file contents are altered. If this requirement is observed strictly, processes are independent of each other and need not come in any particular sequence. The order of operations is dictated purely by the user's convenience. Of course, the result is a rather inflexible file structure; indeed inflexibility of the file structure is the price paid for flexibility in process structures and sequencing. The system overall structure then assumes the form of Fig. 3: the user interacts with various processes, each of which acts on the appropriate portions of a single, formally invariant, data base. The file structure is now inflexible, but the order and manner in which the various processes are invoked may be altered at will. In contrast, the linear structures of Figs. 1 and 2 have an absolutely fixed process sequence but the files change to conform to the requirements of processes. An example chosen from MagNet may illustrate the point: the MagNet file structure does not differentiate between solved and unsolved problems. Both the problem and its solution reside in the same file (there is only one file!). All the solver program does is to replace solution values of low accuracy with values of higher accuracy, and to write into the file a revised accuracy estimate.

Graphics and Distributed Computing

Although character oriented, the hardware of the middle seventies was in principle capable of good interactive computing. But high degrees of interactivity were often not really wanted by remote users of central machines, since communication with an overloaded computer at telephone line speeds left something to be desired. Thus a tendency still per-

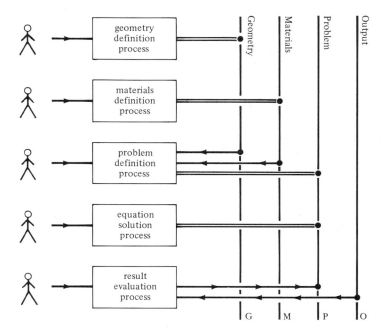

Figure 3. File-oriented software system. A single data base, represented by the four vertical lines, serves all processes. Each process leaves data files structurally intact; processes change file contents, but not form.

sisted to minimize the number of times the computer was accessed and, conversely, to obtain large amounts of printout at every response. To avoid drowning in printout, the obvious answer was to employ graphics terminals and to obtain pictorial or diagrammatic, rather than numeric, output. Early graphics terminals were either of the pen and ink variety, or the storage-tube kind, capable of producing monochrome pictures of very high quality.

A picture is proverbially said to be worth a thousand words. In the world of computer graphics, this estimate is likely to err on the low side; only a drawing of rather modest complexity can be encoded and transmitted to a storage-tube graphics terminal as a string of a few thousand characters. Raster graphics devices require more, rather than less, data. Consequently, fast graphics working speeds could not be achieved. Increased speeds could be achieved, however, by one of two techniques: use of distributed intelligence and use of dedicated computers.

In systems said to have *distributed intelligence*, part of the processing power of the computer is moved to the far end of the communications line. Drawings can then be encoded very compactly, in relatively complicated forms, and reprocessed at the terminal end into simpler commands comprehensible to the graphics hardware. For example, one might transmit in a dozen or so characters "draw an ellipse with semiaxes 0.5

vertical, 1.5 horizontal, centered on the origin", and then translate this instruction locally into hundreds of instructions of the form "draw a straight-line segment to point 0.5070, 0.4910", to form an approximation to an ellipse. The effective communication speed is increased, by making fewer transmitted characters carry more picture information.

When *dedicated computers* are used, the graphics device is moved into the computer proper and made to reside on the main computer bus. The bus speed is generally 100–10000 times the best obtainable on a dedicated serial communications line, and up to 100000 times the speed obtainable on a telephone line. The communication speed is increased in this case by simply eliminating the slowest part of the communication link. This approach has traditionally been part of the minicomputer world, but under the pressure of CAD requirements, even large mainframe machines are now being equipped with graphics devices directly attached to one of the internal buses. In the IBM dialect of English, these devices are referred to as *channel-connected* graphics units.

The general layout of a distributed computing system is shown in Fig. 4. The host central processor and memory, at the left side of the sketch, are tied together by the host computer interconnection bus. (For simplicity, a single bus is shown here although a large computer may have as many as a dozen interconnect buses.) On this bus, which typically operates at a clock rate of 10–20 MHz or higher, information moves at a rate of perhaps 4 million bytes per second (one 16-bit word every fifth clock pulse). The display processor (terminal or local processor) and its local memory are similarly connected, although the processing speeds and data transfer rates there may be slower by perhaps one order of magni-

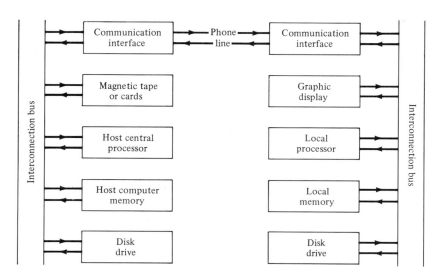

Figure 4. Outline hardware structure of a distributed CAD system. Current systems differ in the relative weight assigned to the host and terminal facilities.

tude. The two communication interfaces are connected by a serial line capable of moving 30–960 bytes per second. This serial connection, which moves data at rates some four magnitudes slower than the host and terminal interconnection buses, forms the true bottleneck in the whole system.

The classical centralized computer facility of the 1970s is identical to the system of Fig. 4, but with very little intelligence at the terminal end. Typically, terminals employed in the seventies contained internal memories of a few words and could understand perhaps half a dozen commands: turn beam on or off, switch from alphabetic to graphic mode, draw a line. The display processor to do so might contain a few chips to implement the necessary rudimentary hard-wired logic. While Fig. 4 still gives a valid description in principle, the left side of the diagram would now contain virtually all the processing and nearly all the equipment used. On the right, only the communication interface and the graphic display would exist in more than vestigial form.

The dedicated minicomputer workstation is also in principle similar to the system of Fig. 4, but tilted to the opposite extreme: the "central" processor has become vestigial, and the terminal processor assumes virtually all computing functions. This situation actually occurs quite frequently; for example, all input/output data preparation and all postprocessing may be done on the local terminal, leaving only the equation-solving task for the central processor. The latter thus handles only one computing process, albeit a very intensive and demanding one.

As a general rule of thumb, man-machine interaction requires a great deal of input/output activity with very short response times, but demands only a little central processor effort. It is therefore well suited to microcomputers and minicomputers, whose processors and memory capacities are usually modest, but which are sufficiently cheap to allow the machine to await the user's pleasure. Conversely, major computational tasks (like solving huge systems of simultaneous equations) require substantial processor activity and large memory but involve very little user communication. They are often best installed on large machines. The art of CAD system design therefore involves two major tasks:

1. Setting up file systems so that processes are essentially independent of each other, yet form natural units in terms of the user's work.
2. Designing processes so that each is run on the most appropriate hardware, avoiding communication bottlenecks.

It is assumed here that the computational tasks to be carried out are themselves well understood, so that system design is the issue, not the invention of mathematical computing processes.

Processes in CAD

Every file stored in computer-readable form must somehow be made man-readable on demand. Furthermore, its content must be communicable to other computer-based systems. The content must be modifiable in a useful and sensible fashion, by people as well as machines. Thus every type of file associated with a CAD system must have associated with it at least one program which might broadly be termed an *editor*: a program that renders the user capable of inserting new items into the file, deleting items selectively, or modifying the file contents. It must have associated with it at least one *file manager*: a program able to create and destroy files of that type, to catalogue the contents of files, and to merge or separate such files. It may also have associated with it one or more *file transformers*, programs which alter the file contents according to some set of rules.

The editing, transformation, and management programs most familiar to the majority of computer users are those employed to work with program listings or other character-based text. For example, a text editor may be used to create a data file containing the coefficient arrays that describe a system of simultaneous equations; a solver program may be used to transform these coefficient arrays into others that describe the solution; and the file management facilities of the local operating system are called into play to catalogue files, delete them when no longer needed, and so on.

Editors

Editors for files in CAD resemble in principle the character editing programs for program development, which exist on every computer system. The latter generally regard a file as being merely a string of ASCII characters and permit operations such as character insertion and deletion. They differ in their level of sophistication and complexity, as well as in their command languages.

The primitive ingredient in a text file is the character. In CAD system files, the primitives tend to be more complex: magnetization curves, finite element meshes, or whole boundary-value problems may be treated as single entities. The editing facilities need not be much more complicated than a typical character editor, indeed they may well be simpler. But they certainly have to be more specialized, because they are dealing with different kinds of quite complicated primitives. For example, "delete this finite element" is a sensible instruction to a mesh editing program, but not to a *B-H* curve editor! It follows that every distinct type of file employed in a CAD system requires a distinct editing program, each possessing its own rules and its own command language.

Editors are invariably interactive programs. Any editor must incorporate its own form of display and must be equipped with its own command language. The display required varies with the file content, but it is generally graphic in CAD systems. The command language may take different forms, but it usually needs to include at least some graphic input facilities. In other words, editing programs require reasonably speedy graphic response, perform a great many input–output operations, and therefore are best implemented in environments where editing program and the user input–output programs are tightly tied together with fast communications. Editing is therefore a task which, in terms of the two-sided system structure of Fig. 4, relies a great deal on local operations and less on the host hardware. The trick in designing good editors is to transfer only very small data sets over the slow serial line (between left and right halves of the diagram in Fig. 4).

File Management

File management functions are those operations which deal with transferring records between files, or in a file, without altering the content of any record.

To be useful, file contents must be easily accessible in every way possible. It must be possible to extract individual records or items and to move them into other files of like type. It must be possible to extract information from individual records and to map it into external forms which can be communicated to other programs. It must be possible to merge files of like kind or to separate them into a larger number of files. Searching, cataloguing, retrieval, copying, renaming, and comparison operations must be feasible. Finally, security arrangements must be acceptable, protecting the files both in the physical and legal senses, i.e., against computer failure and human error as well as against mischief and evil intent.

Curiously enough, computer-readable files probably satisfied all these requirements best during the punch-card era. The punched card is truly a unique medium, being readable by machines as well as by the unaided human eye. As a result, a great deal of file management which now appears very sophisticated was carried out in that long-ago age by simply shuffling decks of cards. Being easily portable, and removable from the computer, security arrangements were not difficult to make either. Unfortunately, magnetic media are readable by machine only, so that even the most trivial operations have to be carried out by programs designed for the task.

File management systems, or file management programs, must provide for all (or reasonably nearly all) required functions. There are two ways of achieving this goal: to write separate file management programs or to structure files so that the general file management utilities provided by any operating system can be used. The latter choice seems the easier at first glance. But closer examination shows up two serious problems— flexibility and portability.

General-purpose file management utilities know nothing of the content or logical structure of a file; they can only consider its physical structure. There is a world of difference between a program which says "these two models differ in the dimension DIAMETR" and one which says "no match word 27 record 106". On the other hand, general-purpose utilities can do almost anything, exactly because they only examine the physical content of a file and do not attempt to interpret it. Their effective use requires a good understanding of the physical file structure and the operating system employed. They are firmly tied to the physical file structure and utility programs of a particular operating system. They are flexible but hard to use and not portable at all.

Special-purpose file managers need know little or nothing about the physical structure of the files being dealt with. They can concentrate on the logical structure of files, leaving the computer operating system to take care of the details of physical format. Their design is tricky, because most files have a richly complex logical structure (as compared with their physical structure); software designers are constantly tempted into assuming "no reasonable person could possibly wish to do *that*". Special-purpose file managers are thus always expensive, can be very flexible and agreeable to use, and are usually highly portable.

File Transformation

A number of processes, whether of interactive or batch nature, can be regarded as transformation operators, which map one file of one prescribed general type into another, generally of some other type. The equation-solving programs included in every computational analysis package are a prime example: they map problem files (which precisely define the mathematical physics problems to be solved) into files of solution values.

In general terms, every transformation may be regarded as mapping some file f, which belongs to a file type F, into another file g of some different type G:

$$g = T f, \quad f \in F, \quad g \in G. \tag{1}$$

Almost all computational processes carried out in CAD may, if desired, be regarded as file transformations.

CAD systems gain flexibility by keeping down the number of intermediate files employed and making all files have direct significance in terms of the user problem. One good method for achieving this objective is to design files so that the sets G and F are the same. In other words, the file structures should be so designed that the input file and output file are the same file; or at least, that they should be structurally indistinguishable. To choose a simple example, the file that contains a boundary-value problem should include in it sufficient space for the solution. The equation-solving program T will then transform between two files f and g which belong to the same structural class F,

$$g = T f, \quad f \in F, \quad g \in F. \tag{2}$$

There is no way to tell, from an examination of the file structure alone, whether the file contains a problem only, or a problem and its solution. (Presence or absence of a useful solution can be recorded in the file itself, by having the solver program set a logical variable to TRUE, or by assigning an estimated accuracy to the solution.) The resulting benefits include (1) one less file type to be managed and (2) automatic inclusion of the data that generated the solution with the solution itself.

Interactive Command Languages

Every interactive computing system must be controlled by means of a command language. Most command languages for CAD systems are semantically fairly simple, but syntactically somewhat more complex than standard programming languages or operating system job control languages. The extra complexity arises from a richer set of linguistic primitives: the conversation between system and user employs both character (verbal or numeric) communication at the keyboard, and graphic or tactile tokens ("this point here") on the graphics screen. Formal as well as classical grammars exist for the former; they are less well developed for the latter.

Command languages used in computer-aided design need to take care of numerous actions, the major ones being:

1. Control of processes: altering the contents (editing) of files, initiating processing steps (e.g., solution of equations).
2. File management: creating, deleting, merging, cataloguing of data files; attaching and detaching files.
3. Display management: specifying what data should be displayed and in what form.
4. System management: identifying passwords, listing available files, signing off.

Command languages are obviously necessary to steer and control these and other activities. Two basic forms of command structure exist, *open-ended* and *hierarchical*. The former is often associated with keyboard input, the latter with screen menus. But the association is coincidental only, for either kind of linguistic structure can be implemented in either fashion, in principle and usually also in practice.

Open-Ended Command Dialogues

Open-ended dialogues between system and user, in their basic form, permit the system to ask only one question: "What next?", and permit the user to issue a single command in response. Any command the system is capable of comprehending will be considered syntactically valid, including (in many systems) commands whose meaning the user has defined himself. The user response may be simple, perhaps even consisting of a single verb, e.g, DELETE. On the other hand, it may be relatively complex, e.g., "delete the current file if there exists a backup copy dated yesterday or later". The command is most often communicated by means of the keyboard.

In open-ended dialogue, it is of course possible for the user to request impossible actions, by issuing commands which are syntactically valid but semantically nonsense. For instance, the request to "delete the current file" makes no sense if there is no file currently open, or if the user has no right to alter the current file. In such circumstances, a system message should inform the user that he is asking for the impossible or the forbidden. In well designed systems, messages of this kind are concise but informative, and above all comprehensible without recourse to the system manuals.

When the user communicates alphabetic or numeric information by way of the keyboard, he normally expects the console screen to echo the input characters. In addition, the system employs the screen to issue system messages, prompts, and responses. It is desirable to let the user identify easily whether any particular message originated with the system, or represents an input echo. The characters employed by the system and by the user are therefore often different. For example, they can be distinguished by color; alternatively user input characters can be echoed in upper case always, while the system is permitted to use upper-case alphabetics only where it is necessary for normal capitalization of words.

Numeric input of data in well designed systems should be convenient and easy. There should be no restrictions regarding placement of numbers on a line, and no distinction should be made between numbers in integer, decimal, or exponential notation, unless of course the distinction is semantically significant. For example, the numeric value *2* ought to be representable as *2* or *2.0* or *2.0E+00* or *2E0* or variations on these. Such vestigial remains of the punch-card era as "type in format F5.2 starting in column 26" rarely indicate clean internal design! Similarly, computers are pretty good at identifying alphabetic characters; "type 1 for yes, 0 for no" is an almost sure sign of a system that wasn't designed, but grew by bits.

Hierarchical Dialogues

Hierarchical dialogues are often used where the syntax of a straightforward keyboard command might become too complicated for the average user to remember. To illustrate, suppose a *B-H* curve is to be drawn, with logarithmic divisions in *H*, linear in *B*. A hypothetical conversation, in which the capital letters denote user commands or responses, lower case the system queries and messages, might run as follows:

```
What next?                            DRAW CURVE
x axis logarithmic or linear?         LOGARITHMIC
How many cycles?                      4
Automatic scaling in x?               NO
Maximum value on x axis?              1E4
```

```
How many ticks per cycle?        3
y axis logarithmic or linear?    LINEAR
Automatic scaling in y?          NO
Maximum value on y axis?         2.2
Minimum value on y axis?         0
Number of ticks on y axis?       22
Full grid or axis ticks only?    TICKS
```

It should be noted that the response LOGARITHMIC is followed by system queries different from those following on LINEAR. In other words, the conversation is tree structured, as shown in Fig. 5. Every time the DRAW CURVE command is issued, the root node of this tree is entered, and some path is pursued through the tree until a leaf is reached. Many parts of the tree may be traversed more than once; for example, the entire tree below the root node has to be gone through twice, once for the x axis and once for the y axis. The user should be unaware of the underlying structure if the system is well designed. The queries should all appear reasonable, and they should all be directed to the issue at hand rather than to solving some internal system problem.

A hierarchical conversation of the sort illustrated can go on and on, next identifying line widths to be used for the curve and the axes, whether to draw with full or dotted lines, whether to erase the screen or to overlay on the previous picture, and so on. Inexperienced users are quite positive about such dialogues, because nothing seems left to chance. They usually cool off rapidly when forced to restart the same conversation for the fourth time, having unfortunately answered the umpteenth question wrong. The experienced system user might well prefer replying to "What next?" with a fairly complete answer, like

```
What next?
DRAW CURVE LOG 4 1E4 LINE 2.2 TICK FULL ERASE
```

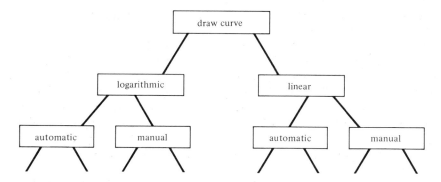

Figure 5. Beginning of a hierarchical command tree, showing the first few branchings. Many different query and answer sequences can result.

provided he can remember the arguments required and their correct order, as well as the default settings assumed for any data not specified. In other words, the experienced user would prefer to work in an open-ended fashion.

There are several compromise techniques which seek to have the best of both worlds. One form, used in PE2D and some parts of MagNet, is the open-ended conversation with prompting. In this form of conversation, the user is allowed to issue commands on an open-ended basis, but any missing sentence elements are not assigned default values automatically. Instead, the system prompts the user to supply additional information, until all required data are in hand. Where this form of communication is used, the experienced user should be provided the option of having the system assign default values instead of issuing prompts. Otherwise, boring and lengthy conversations may result, which merely establish what was obvious in the first place.

A second method for handling the complicated command sentence is the *table edit* approach. All the various data assumed by the system are listed in a table, and the table is displayed instead of any prompts being issued. The user is then given the opportunity to edit the table contents, that is, to alter any parameter or data item listed in the table. To use the above hypothetical example again, the conversation might run

```
What next?   DRAW CURVE

Current settings are:
                         X axis                Y axis
CALIbration =            LOG  4                LOG  3
SCALing     =            AUTO                  AUTO
GRID        =            FULL                  FULL

Which parameter do you wish to change?   _
```

In a full table edit conversation, any user response will result in appropriate alteration of the table entries, followed by a repeated display of the table and the editing query. This process continues until the user replies NONE to the query. Where more than three or four parameters appear in a command, this approach has the advantage of *second-chance capability*: the user can set a parameter to a wrong value, change his mind, and alter it. In a sense, the prompting approach to command sentences is procedure oriented (the user is requested to furnish data for an inflexible sequence of procedures), while the table editing approach is data oriented.

Menu Communication

One traditionally popular form of hierarchical communication has long been the screen menu. There are strong advantages to this form of communication for both user and system designer. Since the designer has full control over the selections that appear in each menu, neither need worry about what might happen in case of an invalid command. Every valid response appears on the menu, every menu item is a valid response, and that's that.

Menus are best used in conjunction with a graphics screen, on which the items available for selection are presented as outlined screen areas. The user then identifies a selection by a graphic cursor hit, lightpen hit, or whatever other form of graphic pointing the hardware system allows. Menu communication can, however, be arranged just as easily, though less elegantly, with a standard alphabetic terminal. For example, the query

```
Length in MILEs, FURLongs, YARDs, FEET,
   or INCHes?  _
```

is a request for menu selection, not functionally different from placing the same five choices on the graphics screen. The only differences are in the form of response, which will be a four-character word typed at the keyboard in one case, a graphics cursor hit in the other.

The great flexibility of menu communication can make it very unwieldy in the hands of all but expert software designers, so it is well to examine menu-driven software with a wary eye. If menu communication is tree structured, it is usually easy to follow the program logic; but it can be difficult to effect any changes of mind (the problem of second thoughts again!) because these may require backtracking through the menu tree backwards. Many menu systems therefore provide (1) a selection, possibly called ACCEPT, to indicate a completed command sequence, (2) a selection, say BACK UP, to allow backing up one menu level, and (3) a selection, perhaps called ABORT, to indicate that the user has become so confused that it seems easier to return to the root of the tree and start over again than to try repairing the damage. While these menu items, and occasionally others, can be very useful, they must be used sparingly. It takes strong user nerves to face with equanimity a menu containing three or four selections that seem similar, like FINISH, RETURN, ESCAPE, ABANDON, TERMINATE, ABORT, QUIT.

In well designed menu systems, only one menu at a time should appear on the screen; otherwise, how is the user to know which of the several menus on the screen is the currently valid one for making selections? Software systems which began life in the storage-tube terminal era often violate this rule, for good technological reasons (it is impossible to erase only part of a storage-tube screen) but with confusion the inevitable result. In systems designed to take advantage of raster-refreshed terminals only one menu appears, but often there is only a poor indication, or none at all, of how the user got to where he is in the menu tree. Taking the example of Fig. 5 again, the user may use a multi-level menu system to build, in effect, a command sentence like DRAW CURVE LINEAR

MANUAL . . . , choosing each word or term from another menu until a full command sentence has been built up. Unless the sentence built to date is actually shown somewhere, it may not be at all obvious what to do with the current menu, and why.

A practical difficulty with menus lies in the use of screen space. The total information content of a 512×512 resolution graphics screen is one-quarter megabit, no more and no less; any screen space used up in menu presentation is effectively subtracted from space available for graphics. One solution to this problem is to reserve no menu strip or menu area on the screen, but to overwrite any arbitrary part of the graphic display with the menu; once the menu has been satisfied, the screen can be restored to its initial state. Again, storage-tube hardware is unsuitable for this solution.

For the graphic displays needed in design, a square screen space is generally best, since objects to be designed might be of any shape at all. Communication areas ought therefore be so placed as to make the available picture space essentially square. If a conventional low and wide screen is used (the so-called *landscape* screen format), the full screen height ought to be used for pictures, and a vertical strip down one side for commands. If, on the other hand, a tall and narrow screen is employed (*portrait* format), a horizontal strip at the top or bottom is to be preferred, with the graphics area occupying full width.

Programmability of CAD Systems

In some CAD systems, the commands that control the system can be accumulated into files and re-used. It is fair to view these command sequences as *programs* for executing certain actions, so that the CAD system ceases being a tool for direct problem solving and becomes instead an interpreter for a very high-level programming language aimed specifically at magnetics. Systems vary in their programmability, from rudimentary to very sophisticated.

Stored Command Files

CAD software systems have traditionally been classified into *batch* and *interactive* types. From a formal point of view, the distinction is one of asking from where the stream of instructions comes which is to guide the system. Both batch-run CAD software, in which all instructions must reside in a file at the start of the run, and interactive software, in which the user may issue instructions as the analysis proceeds, have their advantages, so much so that in recent years the trend has been toward blending their characteristics through programmability.

Batch-run analysis systems have an advantage over their interactive counterparts in one interesting respect. All commands and data needed to perform a particular task must first be placed in one or more files of well defined form. These are then read and executed with no user intervention during the program run. Should alterations to the data, or minor changes in the processing steps, be required, the user need not re-create the entire input file; it is only necessary to edit that file to remove or include the modifications desired. In other words, an analysis can be repeated (possibly with different data), or another one of substantially the same sort can be carried out tomorrow, by making a copy of the command and data files of the old problem and altering them sufficiently to cover the new problem. No machine resources are saved in processing the altered problem, but the human effort required to define it is minimal. To be very precise, the input file or files communicated to the analysis system ought to be viewed as a *program*, written in a highly magnetics-oriented language, and the analysis system could be thought of as an interpreter for that language.

Interactive analysis systems are often better liked by users, particularly novice users, because of their flexibility. However, they take their input from keyboard and graphic stylus or mouse, not from a file. On gaining familiarity with any interactive system, it quickly becomes obvious that for similar tasks, similar command sequences are executed time and again. It makes good sense, therefore, to allow command sequences to be stored and filed, for re-use as occasion demands. Doing so permits the interactive system to be run in partially batch mode. Storage of command sequences in this way is of course hardly a novel concept; classical batch-processing systems offer it automatically since they can operate in no other way.

Stored command files can be created and incorporated in interactive systems in two distinct ways, either by having provision for them within the CAD system itself or by making use of the operating system under which the CAD software is run. Working from the outside is certainly the easier choice for the implementor, for many operating systems permit users to create and employ command files without any reference to the program itself. For example, the Unix operating system, as well as several others, allows peripheral device reassignment, so that keyboard input—or what the CAD program thinks is keyboard input—can in fact be taken from a file. To run commands from a file under such an operating system, no modification at all is needed to the CAD software, and any CAD system whatever can be made to function in this mode. However, there is usually a restriction on graphic input. Most interactive CAD systems actually use two independent input devices, the keyboard and the stylus (or mouse). Reassigning the keyboard input to come from a file instead does not reassign the mouse; thus two input files, and two device reassignments, are required to substitute fully for terminal interaction.

Keyboard conversations in most interactive CAD systems are reasonably comprehensible and straightforward, so that the preparation of command files is easy enough. On the other hand, graphics tablets and mice work by transmitting stylus or mouse coordinates, encoded as strings of characters, and these are not nearly so easy to prepare independently.

Knowing where to point with stylus or mouse amounts to knowing where on the screen (in terms of absolute screen position) to point, and how to do the position encoding, a task perfectly possible to carry out, but not particularly agreeable nor one likely to be error-free.

Under the Unix operating system, and under a few others, the difficulties of device reassignment and of command file creation can be overcome in a clever manner. Unix includes a mechanism called a *tee*, whose name derives from analogy with plumbing: a data flow can be intercepted, with a copy of the data trickling to a secondary destination. Using this mechanism, the user can intercept and record in files both keyboard and graphics device communications, while executing an actual problem run. For a subsequent rerun, the two files can then be replayed, possibly with some alterations.

Heuristics and Formal Programming

Interactive CAD systems can be driven from command files and thereby can be made to act like batch systems. But as may be evident from the discussion above, there are two fundamentally distinct ways of creating the command sequences: *command file editing* and *heuristic learning*. The former closely resembles programming in well-known languages like Fortran or Pascal: the desired sequence of commands is written into a file, the file is given a name, and the command sequence thus filed is executed when requested. In contrast, *heuristic learning* requires no planning or separate *programming* phase, at least not visibly. In its simplest form, as given above, it consists merely of recording the sequence of commands given by a human user, for replay on another occasion.

Heuristic programming is possible by using operating system facilities, but is better done from within the CAD software. It is accomplished by asking the system to *learn* or *remember* commands as they are actually entered and executed at the terminal in a real working session. The difference between using the operating system facilities (where they exist) or the CAD system itself becomes evident in two ways. To begin, the user need not know much about how to use the operating system, an advantage if there is no other reason to be interested. More importantly, the files as maintained by the CAD system need not contain direct character images of terminal communications, as would be the case for files maintained by the operating system. This advantage can be of major importance with records of graphic inputs, for the CAD system can maintain these in a form of much greater significance to the user. For example, cursor selections from a menu can be recorded by remembering the menu item selected rather than the screen coordinates of the cursor hit. Should it be necessary to edit the command file later on, this approach can simplify life a great deal.

Both the heuristic and the formal programming approach to creating program files have their adherents, and good arguments can be presented for both. In reality, the differences are not quite so great as might seem.

The key idea of heuristics is that the remembering is "natural"; the user need take no special steps as would be needed in formal programming. However, most heuristic systems will remember false starts, missteps, and error corrections along with everything else, so at least one or two trial runs will be made by users in real circumstances, before committing a working sequence to file; and if the steps are noted down so as to get them right, the process has in fact become one very similar to program writing.

Although the construction of command files can make an interactive system act much like one designed for batch operation, the converse is not necessarily true. In fact, it may be said that the better the batch program, the more difficult it is to turn it into an interactive one. The reason is that well designed modern batch software tends increasingly to use *declarative* or *descriptive* command languages, while interactive systems tend to be *imperative* or *algorithmic*. The difference is rather like that between, say, the Prolog and Fortran languages. One attempts to describe a logical structure, which may eventually imply what processing steps are necessary for a given task; the other prescribes a sequence of actions, which may eventually imply what data must be available in order for the actions to be carried out. Because the statements in a descriptive language can be made relatively context-independent, modern batch-oriented command languages allow the user a great deal of freedom in the order of place-ment of commands. As a simple example, it makes no difference in the MAGGY system whether a particular material is described before or after it has been used in a device model; it is only necessary for a full description to exist at the time it is actually required for computation. In a sense, the batch system can rearrange the commands into whatever sequence it likes, because it will have them all in hand before problem analysis actually begins; the interactive system finds it much harder to afford this luxury.

Programs and Command Files

Although the writing of command files is often referred to as "program-ming", the terms *program* and *programming language* are usually thought to have a more strict meaning. Programs, in the stricter sense, are com-mand sequences capable of taking different actions, based on logical deci-sions arrived at through examining data. In other words, some kind of branching mechanism is implied, which may have the form of an *if . . . then . . . else . . .* construct, but may take some more primitive shape. Various constructs are familiar from conventional programming languages.

The minimal mechanism possible for taking logical decisions is a single logical variable whose value may be examined. It is not even necessary for this variable to exist as an explicit named entity, nor for any explicit mechanism for setting its values. Assembler language programmers will be familiar with the notion of a *zero flag*, set if the last operation to be executed produced a zero result. Similarly, Snobol programmers feel quite at home with a *success flag*, a logical variable which is assigned a value in accordance with whether a previous action was successfully carried out. Program flow can be controlled by such a simple decision mechanism, provided it is augmented with two other inventions: some form of *branch* instruction, and a *tag* or *delimiter* which permits identifying a set of command statements. Given these three facilities, it is possible to make statements which say, in effect, "*if* the flag is set, *then and only then* execute the following statements, up to the next tagged one".

Fully flexible programmability makes hierarchical forms of structure desirable. One example of such a structure is MagNet, where any acceptable sequence of commands may be turned into a new (user-defined) command, which can then be used in its turn as a constituent component of other programs. Several examples of such constructs will be found above, in connection with the MagNet mesh generator and postprocessor structure descriptions.

Even if the programming language is very closely keyed to the problem area at hand, few people are actually capable of writing more than a few lines of program without likelihood of error. Any fully programmable system which permits data-dependent command sequencing should therefore also include at least a simple form of debugging facility, such as the ability to execute the program one step at a time.

Workstation Layout

The physical design and layout of a CAD workstation has a very strong bearing on user acceptance of the entire system. The choice of graphics devices, design of display screen layout, and the physical positioning of the operator all assume importance to people who intend to spend extended periods working with the system. Computing may not always cause a pain in the lower back—but having the keyboard three inches too high above chair level is almost guaranteed to do so!

Graphic Input Devices

Graphic input in CAD systems takes two basic forms. A great deal of communication between user and system takes place through the graphic display—the user identifies elements of the display or selects choices from menus, by pointing at them. This type of input requires sufficiently rapid response to allow repositioning a crosshair cursor or other marker on the screen every 15–20 milliseconds, so as to follow hand movement. Extreme precision is not required, since the device is generally positioned freehand. A second form of graphic input occurs in applications such as the digitizing of curves or graphs, where essentially numeric information must be transmitted. Precision is essential in this case, but speed is less critical.

The graphic input device of choice today, incorporated into the majority of CAD systems, is the digitizer tablet. The tablet is used with a stylus resembling an ordinary pencil, though some workers prefer a "puck" which lies flat on the tablet. Either device contains a small sensing coil which determines its position on the tablet and communicates it to the outside world through a cable. A position signal is sent out every few milliseconds, so that the stylus can be easily used to follow ordinary manual sketching. The most common varieties of tablet have an active surface about 30 cm square, a little larger than an ordinary writing pad, with a spatial resolution better than 0.2 mm. Larger sizes exist, up to about two meters square. They are sometimes used for digitizing existing drawings or maps. The tablet has won wide acceptance because its precision is sufficient for digitizing graphs while its speed is quite high enough for cursor steering.

Other graphic input devices encountered occasionally include joysticks, thumbwheels, trackballs, lightpens, and mice. They all have the disadvantage of not being usable for numeric digitizing work. The mouse is a convenient and widely recognized device for cursor steering. The other devices range from good to just barely acceptable.

Joysticks, thumbwheels, and trackballs all have two rotary position detectors—potentiometers, position encoders, or the like—which detect x and y positions independently. Thumbwheels have been used for years in storage-tube terminals. They have the advantage, much liked by some, of permitting cursor movement in the x and y directions independently. In fact, they are the only device with which it is possible to move the cursor along a true vertical or true horizontal line. Joysticks come in two varieties, position controlling and velocity controlling. Position control joysticks are quite agreeable, but cost as much as a graphics tablet, since two-channel analog to digital conversion with at least 9- or 10-bit precision is required in both cases. Velocity control joysticks are cheap to make, indeed they even exist in the extreme form of four pushbuttons which move the cursor at constant speed in one of the four compass directions. They seem to be good enough for video games; but they have been so roundly condemned by the CAD user community that very few systems even offer them as options.

When the first refresh graphics terminals were announced in the 1960s, they used the lightpen as an input device. In principle, the lightpen seems attractive; in practice, it is not. Because it is used to point directly at the display screen, it requires working on a vertical surface, which is tiring. Alternatively, the display tube must be mounted horizontally, so the operator has to work at a table with no kneeroom at all. Sufficient positional accuracy can be achieved with lightpens only at the blue end of the spectrum, a matter of no great consequence in monochrome but rather disturbing in color work where every lightpen hit is accompanied by a blue flash of the screen. Lightpens cannot be used to digitize existing drawings, cost as much as tablets, and seem to offer no compensating advantages.

Ergonomics of Two-Screen Working

Many users of CAD systems prefer working with two screens: one graphics display screen and one alphabetic display. Keeping the command conversation off the graphic display means that the graphic display is not cluttered with commands, yet the last one or two dozen commands remain visible on the alphabetic ("console") terminal, so the user can always see how he got to where he is. Working with two screens is quite agreeable to most people, provided the work station layout is appropriate and the software systems keep the needs of two-screen working clearly in mind. Both requirements are primarily ergonomic, and both are tricky to meet satisfactorily.

Most people seem to like an eye-to-screen distance such that the screen subtends a lateral angle of vision around 15–20°. Two screens side by side, each at the appropriate viewing distance for a 15-degree angle of view, require turning the direction of vision (combined neck and eye movement) through about 20°, judged by many to be an inconveniently large lateral movement. The problem is at least partially curable by placing the graphics screen above the alphabetic screen. An upward eye movement of some 10–15 degrees seems to be less bothersome than the corresponding lateral movement. Movement is also reduced, since display screens are wider than they are high.

Cursor steering with a graphics tablet, using a stylus or puck, is a very natural movement. Indeed when using a stylus it is easy to forget that one is not actually writing on the screen—provided the tablet and screen are well aligned. The tablet top and bottom must be parallel to the horizontal edges of the display screen to within five degrees or so. If the misalignment is much greater, hand–eye coordination becomes very difficult. It is quite helpful, but not essential, to have the tablet directly in front of the display screen. It is also helpful to have the screen and tablet subtend the same perceived angular width; if the tablet is immediately in front of the screen, the tablet area and screen area should be similar. However, it is easy to adapt to different size scaling, at least within a factor of 2 or 3 in area.

In general, there are two unalterable leading dimensions to a workstation. The width of the keyboard, which is dictated by the size of the human hand, is always about 35–40 cm (though wider ones exist). The size of the graphics tablet is usually about 35–40 cm square, to accommodate conventional paper sizes, which are also related to the size of the human hand. The workstation thus has to have a relatively inflexible width of about 75 cm or wider, with the keyboard and tablet side by side, and next to the corresponding display screens.

The width of a normal graphics workstation is large enough to make continual switching from keyboard to tablet and back rather inconvenient. Good software design therefore requires hand movement from keyboard to graphics device, and eye movement from one screen to the other, as rarely as possible. For example, where the system designer has a choice between a keyboard query and a menu, he should choose to employ the device likely to have been used most recently. Questions interspersed in strings of graphic input should therefore expect replies via menu selections on the graphics screen, while questions embedded in numeric input should require keyboard replies. Which device is active, should be evident at all times by glancing at the screens. The graphics cursor and menus on the graphic screen should appear only if graphic input is actually expected; conversely, the alphabetic cursor should appear on the screen only when input is awaited. Above all, the situation should never arise where the user and the system both wait for the other to speak first!

Design considerations of importance include such apparently mundane matters as the height of the screen and keyboard. It is surprising how many CAD system users develop physical tiredness and "computer fatigue", simply because their terminal keyboards are installed on tables designed for writing, and therefore three inches too high for comfortable keyboard use!

Color or Monochrome?

In the commercial world of CAD systems, the axiom is that color sells.

Color displays are indeed very attractive, and they really do convey more information (provided the software is built to make good use of color). Although color displays are more expensive than monochrome, the display subsystem typically accounts for a good deal less than half the total hardware cost of a CAD system, and hardware generally costs less than the CAD software associated with it. In overall system cost, color therefore comes perhaps 10% higher than black and white, thus meriting serious consideration.

If color really does do more, what objections are there to it? Three, so far as one can at present tell: (1) availability of hard copy, (2) reproducibility, and (3) resolution.

Paper copies of monochrome displays are readily available from electrophotographic hard-copy units. For color displays, there are two main technologies available: photographic reproduction and ink-jet printing. Color slides are a cheap form of hard copy, since a truly excellent 35 mm camera costs less than a replacement picture tube for the television monitor! Slides are a superb medium for presentations, meetings, and conferences, but not nearly so useful for actually consulting in files or for inclusion in a report. Eight by ten inch instant color prints can be made, or less spectacular smaller ones, but at a high price. Not only are they expensive, but they are almost impossible to reproduce, so that fifty original prints must be made for a report to be circulated in fifty copies. Ink-jet printers are a bit expensive and somewhat fussy to maintain, even though they are improving rapidly on both counts.

Reproduction of color hard copy may in fact be the worst of all problems with color. Many engineering reports are circulated to five, ten, or fifty people, numbers too small to justify color separation printing processes yet too large to print duplicate originals. Worse yet, many internal reports are printed on demand, by xerographic processes. Color xerography, like the lightpen, has long been heralded as the technology of tomorrow; but in most offices tomorrow has yet to come. The impact and communication value of color are therefore often lost, so many realistic CAD system users employ color for making slides, but incorporate into printed reports only pictures in which color is not a vitally important ingredient.

Because color television technology differs from its black and white counterpart, color monitors of comparable resolution are costlier and also fussier to maintain. At the present time, 512×512 color systems are common, and 512×768 is available routinely; 1024×1024, which was considered difficult and fussy only a few years ago, is no longer unusual. Monochrome systems of 1024×1024 resolution are available at quite acceptable prices. Both color and monochrome systems of 2048×2048 pixel resolution can be found, but they are still difficult to maintain and care for, as well as initially expensive. At resolution levels up to about 512×768, monochrome displays enjoy a distinct price advantage over color because they make use of conventional television technology; for higher resolutions, however, the line standards, scan rates, and other details depart in any case from those employed for the living-room TV set.

Interactive CAD Systems for Magnetics

There now exist quite a few CAD systems capable of solving problems in magnetics and electromagnetics. Many have been reported on in the technical literature but are not generally available for proprietary or other reasons. Four packaged interactive systems for two-dimensional analysis and two batch-run programs are now available in the marketplace. The interactive systems are:

AOS-Magnetic, written by A. O. Smith Corporation,
FLUX2D written by Grenoble Polytechnic Institute,
MagNet written by Infolytica Corporation,
PE2D written by Rutherford Laboratory.

These four are all strongly oriented to magnetics. They will be described in some detail below. There do exist also two batch-oriented systems that merit consideration; they are treated separately in the next section.

Hardware Systems

The four interactive systems now available span the range of possible hardware systems, from the time-shared monster computer with totally dumb terminal to the dedicated microprocessor operating autonomously. While they are now available in versions intended for various kinds of computers, the structure and layout of every system are strongly influenced by its history. Interestingly, no two were intended for identical hardware initially, although they are growing together with increasing hardware standardization.

AOS-Magnetic and FLUX2D were both written to run on central computers. Initially, there existed only one installed copy of each, and users dialled in over telephone lines. The precise nature of the distant large-scale computer is generally of no interest to the user, indeed it might easily be changed without his being aware of it. To the user, the distant computer appears only in the form of an operating system, whose command idiosyncrasies and file structures must be mastered (at least on a rudimentary level) so as to use the system. In this respect, FLUX2D is rather pleasant, for it was designed to run under the MULTICS operating system which most users quite like, even though it is now considered obsolete. Communication speeds are of course restricted by the telephone line. But since users dial in under conventional communication protocol, the line speed—usually from 300 to 1200 baud—can be selected by the user when starting work.

AOS-Magnetic can be used with a truly rock-bottom terminal hardware configuration: a Tektronix 4014 (or 4010) storage-tube terminal and a telephone coupler. There is only the single display screen, so both the command conversation and all graphic output must share screen space. Since a storage tube is employed, no selective erasure is possible, all one can do is erase the entire screen and start over. At telephone line communication speeds, some inconvenience in working patterns must be accepted. Recent developments in AOS-Magnetic include a version to run on single-user workstation computers, as well as a version to use raster graphics devices. However, AOS-Magnetic makes little real use of the high degree of interactivity possible with these very powerful hardware systems. Because its history began with nonerasable graphics and slow communications over telephone lines, AOS-Magnetic uses graphics more for *display* than for *communication* and keeps command dialogues brief. Much of the communication is therefore keyboard oriented; the user is obliged to read numbers off the screen and type them back in at the keyboard, because it is not possible to point at screen objects to identify them. The minimum hardware cost to run AOS-Magnetic is the price of a good terminal and a telephone coupler.

FLUX2D also employs a storage-tube display, but in a two-screen configuration. The second screen is a conventional "glass teletype" terminal. A single telephone line is used. Output destined to the two screens is switched by a local controller, a special-purpose device which is instructed by FLUX2D whether any particular output stream should be sent to the alphabetic or graphic screen device. No selective erasure is possible, but fewer full screen erasures are needed than with AOS-Magnetic, since all command conversation bypasses the graphics screen so there is less screen clutter. The local hardware setup thus comprises one storage-tube terminal, one conventional communications terminal, a telephone coupler, and the output controller. Minimum costs are as for AOS-Magnetic, plus a simple alphanumeric terminal and the switching controller. Recently, FLUX2D has also become available in a single-screen version, with some screen crowding the result of handling both the graphic and alphanumeric data streams.

MagNet represents the opposite end of the design spectrum, having been initially built for a highly intelligent terminal workstation loosely (and optionally) coupled to a central computer. The workstation contains a microprocessor and rapid graphics display. All computations can be carried out within the workstation, which in normal operation can handle all the required work except perhaps for the very largest arithmetic processing tasks. MagNet is the only one of the four systems in which the user does not normally time-share all computing resources, including the input-output processing facilities. Recent developments include versions of MagNet intended for intelligent raster-refresh color terminals driven by a superminicomputer. Such hardware configurations permit greater degrees of time sharing, thus slow down interactivity if there are many users on the system, but improve response time as compared to a pure small workstation environment. MagNet cannot be used with storage-tube terminals, because it was originally designed for raster graphics use and makes frequent use of selective screen erasure. Operating systems under which MagNet runs include Unix, all current VAX/VMS versions, and others. Minimum hardware costs are those for an intelligent raster graphics terminal.

PE2D occupies a position halfway between the hardware extremes: it ordinarily runs on large minicomputers or superminis (Prime 750 or VAX-11/780) and communicates with the user through a single keyboard and screen. Implementations on IBM and Amdahl hardware also exist. Time-shared operation is the norm, and the computer operating system (PRIMOS or VAX/VMS) is quite visible to the user. Communication takes place over serial lines, which may be purely local and dedicated. Typical line speeds are 300 or 1200 baud for remote (telephone line) operation, 4800 or 9600 baud for users located at the computer site. The use of a single screen does lead to screen clutter, but the relatively fast communication speeds in local operation imply that full erasure and replotting is much more acceptable locally than when working over telephone lines. The graphics terminal to be employed is not rigidly specified. Its original versions employed storage-tube terminals, and some aspects of system structure bear witness to that fact. Both storage-tube and refresh graphics terminals can be used with current versions, since all graphic input and output operations are routed via the GINO-F graphics package. GINO-F provides at least nominal device independence, which unfortunately means that, most of the time, all devices are reduced to the characteristics of a storage-tube terminal.

Use of these systems with terminals different from the above is in principle possible, provided a terminal driver package can be obtained to interface the CAD software to the graphics device. The range of available devices is moderately narrow for FLUX2D and AOS-Magnetic, since the PLOT-10 proprietary graphics package is used for displays. PE2D communicates with all graphics devices through the GINO-F package, which supports a fairly wide range of terminal hardware. MagNet employs the GKS international graphics standard, so that terminal driver software is likely to be available for a variety of terminals for quite some time to come.

Methods and Problem Range

All the interactive systems described here use the finite element method, and all provide at least first-order triangular elements. FLUX2D also includes quadrilateral elements, thereby introducing some extra geometric flexibility. PE2D makes provision for second-order elements in the problem definition and solution phases, but restricts local material property variation in the postprocessing phase. Second-order elements can prove particularly useful for linear problems.

The total number N of nodal variables allowed depends on hardware limitations, and on the equation solving methods used. AOS-Magnetic and FLUX2D both employ a modified Gaussian elimination method, in which the storage requirement is roughly proportional to the three-halves power of N. However, both systems are normally installed on relatively large computers and therefore can handle a substantial number of nodes. Indeed, the number of nodes is likely to be limited by the inconvenience of data input and output over a slow communications line, rather than by the program itself; some users seem to feel that AOS-Magnetic is rather tiresome when more than 400–500 nodes are used, even though the program capacity is much larger. FLUX2d is limited by the same considerations, but provides better accuracy for the same number of nodes N, because second-order quadrilateral elements are employed. On a basis of accuracy for a given problem, N nodes used by FLUX2D can therefore be worth $2N$ or occasionally even $4N$ nodes used by AOS-Magnetic.

Both MagNet and PE2D use preconditioned conjugate gradient solvers, in which storage requirements rise only linearly with node count. They are therefore capable of coping with comparatively large numbers of nodes even when using modest engineering workstations. When running on local computers, there is not much difficulty about communication line speeds, so that the available maximum node count can be used fully. On the other hand, both run on comparatively small computers. Hence the maximum usable node count lies in the range 1000–5000 for these systems; it is of course machine and operating system dependent. Equivalences between systems are in any case hard to establish by node counting since minor differences in the input mesh generators cause users to employ different working patterns, thus arriving at different node counts for similar problems.

All four systems are capable of solving one-component magnetic vector potential problems in the x-y plane, with soft saturable magnetic materials and linear materials. True permanent magnet representations are available, but even in their absence, all four systems are capable of solving many practical problems, approximating remanent magnetization by equivalent distributed currents. All four can deal with scalar potential formulations also; however, some precision difficulties may arise in the modelling of magnetic material properties in the saturable scalar potential case. Solution of axisymmetric (r-z plane) saturable material problems, which requires little beyond the replacement of x-y plane finite elements by r-z plane finite elements, is available in all four. Similarly, the solution of linear time-harmonic (eddy current) problems requires replacing the real, nonlinear mathematical subprograms by complex, linear ones; again, this possibility is available in every case. Axi-periodic elements, on the other hand, are not invariably available. Nonlinear time-varying problems (transients) are much more difficult, because a sequence of solutions must be computed, one for every time step. The resulting quantities of data are huge, and although such problems can be solved by PE2D in certain cases, suitable postprocessing facilities are not even in early prospect for the four systems described here.

Between 70% and 90% of the program code in all four systems is devoted to data management, the rest to mathematical operations. Generally, the better the interactive response, the larger is the proportion of program code devoted to data handling. But since all systems must necessarily contain the same mathematical core, a larger proportion of code devoted to data handling automatically implies a larger total amount of code and a more expensive package!

Procedure or File Orientation

In procedure-oriented systems, the user is generally encouraged to choose the path he wishes to follow at an early stage. He is then locked in to that particular path, and cannot change his mind very much, except by restarting. File-oriented systems give the user much more of a second chance, but they can confuse the inexperienced precisely because there is a lot of choice at every step. Large amounts of disk activity, and relatively few input–output operations over the communications line to the terminal, result from a procedural orientation. Large computers, with fast disk access and slow serial lines, therefore encourage system designers to pursue a procedure-oriented direction. It is therefore not surprising that the systems which most nearly resemble classical batch processors are most strongly process oriented. Conversely, those which assume dedicated hardware and support software tend to have a stronger orientation to files.

AOS-Magnetic and FLUX2D, as one might expect, present the user with strongly sequential process steps. There is an advantage for the novice user: with but one road to follow, there is little to get confused about. The main disadvantage is that many files are generated, which are then reprocessed and deleted, the newly generated files at each step containing the necessary information from their predecessors. Unfortunately, such a structure means that an upstream mistake cannot be rectified except by restarting, and that information related to some aspect of one problem cannot be readily moved into another problem run. Files are generally not easy to extract from either system, both because few provisions for file transfer exist and because of the slowness of communication. FLUX2D has an edge in this respect, however, because its storage forms keep file sizes relatively small.

PE2D lies at an intermediate position in hardware, and therefore also in its structural orientation. File formats are fewer than in the centrally run programs, and most files are retained in a reusable form. However, many parts of the data are kept in intermediate forms hard to re-use without extensive processing. For example, the finite element meshes employed in analysis are stored in a compact form by regions, and a mesh generator program is run whenever the explicit mesh is actually required. The file formats are well chosen, however, and many of the general-purpose system utilities provided by the computer operating system can be used to effect file management. File extraction is fairly easy, on magnetic tape or disk packs, since users are generally located in close proximity to the computer. The analysis processes are coupled, but quite loosely.

MagNet is strongly file oriented; few distinct types of file are used by the system, and each is externally accessible through file management utilities. The procedures that constitute the system are entirely uncoupled, and communicate only through the files.

Programmability of a very rudimentary sort is provided in AOS-Magnetic and FLUX2D, in that data files can be retained and read back. Of course, command programming through the operating system (and device reassignment) is always possible. At the other extreme, MagNet is programmable, without recourse to the operating system, in the fullest sense. It provides conditional execution and program branching, allows symbolic parameter passing between user-defined programs, and provides modest debugging aids.

Access and Maintenance

The mode of user access allowed by each of the four systems described here is suited to its hardware implementation: they vary from rental by the hour to outright purchase.

The two centrally installed systems are available on a rental basis, AOS-Magnetic in North America and FLUX2D in Europe. In both cases, there is a reasonably moderate annual subscriber fee to be paid. Use fees (plus telephone bills) are likely to exceed the annual subscription by a large margin, except for very occasional users. Rental fees are levied through rather complex charging arrangements, different schemes being attractive to different customers. As a very rough guideline, the average user probably pays the computing bill, plus a program use royalty of about 100%, and the annual subscription fee amounts to an hour or two's use. Users do not have access to the system source code, nor should they need any. There is no problem about system maintenance, since in each case there exists only a single centrally located and maintained copy of the system.

When a system is purchased for installation on a user-operated computer, matters are of course a little different. A long-term rental fee is then paid, and no hourly or other user charges are levied. Fees range from perhaps double the hardware cost for small systems, to considerably less than half the hardware cost for large minicomputer installations, typically in the region of $10000–$50000. In point of legal fact, what is purchased is not the system, but rather the license to use one for a stated period of time, ranging from a few months to 19 years. Obviously, license fees depend in part on the length of lease. Programs are normally licensed to be run on a single computer. Maintenance is provided, for a fee, by the supplier. Installation is usually performed by the vendor and does not normally cause major problems; users receive machine code suitable for the computer in question. In most cases, initial user training is included in the license fee, the amount of training being obviously more ambitious with the higher-priced systems.

AOS-Magnetic is supplied by its originating organization and is maintained by its originators within North America. FLUX2D is supplied by CEDRAT-Recherches, a commercial consulting firm. Maintenance is provided by the French university group which originated the software, through a contract arrangement with the the vendor. Maintenance is currently available for FLUX2D within Europe.

MagNet runs only on user-operated computers and is therefore available by purchase of a long-term license. As in the case of AOS-Magnetic, the usual term of the license is long (from 5 to 19 years), and falls short of perpetuity only for legal reasons. Prices vary considerably with the hardware configuration and the software options selected. The normal license agreement covers maintenance only for an initial period of a few months or a year; thereafter, annual maintenance contracts are available. Maintenance for most systems costs about 10% of the current system price for large systems, and somewhat more for small ones. Individual portions of the system (e.g., mesh generation only) are available. Maintenance is available within North America and Europe, by the originators and developers of the software.

PE2D is available on a lease basis, the normal rental contract being for three years. Few if any parts of the system are usable individually, the system is too tightly integrated for that. PE2Dis normally delivered as object modules for a particular user operating system; it is maintained under arrangements broadly similar to those applicable to MagNet. The commercial history of PE2D is colorful, the system having been delivered by various different vendors at different times. It originated with the Rutherford High Energy Laboratory, and proprietary rights are now held by the British Technology Group, who hold the rights to most scientific software developed in UK government laboratories. BTG do not sell software, however; they license it for others to market and distribute. Infolytica Corporation hold a license to provide marketing and user services for PE2Din North America; European users are served by Vector Fields Ltd.

In general, there seem to be three feasible solutions to the problem of maintenance, and all three are illustrated, more or less, by the available systems. Centrally installed systems are centrally maintained, since the same version of the program becomes available to all users automatically. Systems installed at the user site can come in two stable forms: those delivered in core-image form, which cannot be modified by the user, and those delivered as source code, which the user is free to modify to suit his own desires. One can reasonably expect the vendor to guarantee only the unmodified product, so that full maintenance and guarantees of performance imply distribution of no source code, only executable core image. Conversely, distribution of source code implies minor maintenance at best, for no system vendor can possibly undertake to guarantee software modified by the customer. One cannot realistically expect a software vendor to guarantee programs "improved" by someone else—no more than one expects an automaker to back his guarantee after the local speed shop has altered the suspension and rebored the engine!

Documentation

The world's most perfect software is useless until sufficient documentation is provided to inform the user of the facilities provided by the system and how to use them. Ideally, at least three levels of documentation should be provided for every system:

1 *Aides-memoire to the user.* A short manual is useful to jog the experienced user's memory on how to do things. Such documents are often called "reference cards", even though the reference card for one popular operating system runs to 18 pages!

2 *Full user documentation,* which precisely defines every facility in the system, and defines the expected action in response to every command. This document is much more important than is often realized, for it is this document which defines just what the vendor has guaranteed.

3 *Maintenance documentation.* It is not enough to furnish source listings; the design principles should be embodies in an appropriate document. Subroutine calling graphs, details of common block structures, and interrelationships of arry sizes, for example, should be made clear independently of the listings.

Reference cards are often best embodied in the system itself, so that quick reference may be made to system documentation at the user terminal, without consulting a separate document. This approach works best in local workstation systems; the response of distant mainframe machines can be frustratingly slow. The full user manuals should be furnished in paper form, preferably in more than one copy so that an archival copy can be stored away in a safe place. Full user manuals for any system need to satisfy two requirements, which often conflict. They must define precisely what the system can be expected to do; they often also try to be tutorial in nature, to help the user get acquainted with the system. It is very difficult to be expository and definitive at the same time, and few writers of system documentation succeed.

Where full maintenance is provided, and no source listings are available to the user, maintenance documentation can of course be abbreviated. It suffices then to give details of what assumptions the hardware and the operating system software must satisfy. In extreme cases, the maintenance documentation might consist of no more than a telephone number!

AOS-Magnetic documentation is generally expository rather than definitive. Enough examples are provided to allow novice users to get started. Less provision is made for experienced users. No maintenance documentation is provided for users dialling in to the centrally maintained system. Training courses for users are available.

FLUX2D is well supported by a manual of more than 100 pages; Mag-Net comes with a thick book of some 600 pages. The documentation is adequate for tutorial purposes but is primarily definitive. For FLUX2D, a good reference card is provided; MagNet uses on-line reference card files, called up by a HELP command. A hardware (and for MagNet, operating system software) brochure is available in each case. No software system maintenance documentation is provided, since users cannot modify the source code in any case. At the time of writing, FLUX2D documentation is available in the French language only; an English translation appears to be in preparation.

PE2D is supported by an essentially tutorial user manual of about 100 pages, by a printed reference card, and by a reasonably extensive on-line reference card system. However, the on-line reference cards are accessed by commands which are not uniformly the same, so it is first necessary to learn how to get to them. There is little maintenance documentation available, but none should be needed by users possessing an object code license.

Batch-Run CAD Systems

In addition to the interactive systems described above, two batch-run programs are currently available which merit consideration for industrial work. They are

MAGGY written by Philips N. V.
TWODEPEP written by IMSL, Inc.

Batch-run programs are much less convenient in use than interactive ones, but they have compensating advantages. They generally run on almost any medium to large computer, requiring no special facilities and demanding little by way of system loading, time availability, or interactivity. They are therefore very convenient for occasional users.

The two available systems differ considerably. MAGGY is keyed to magnetics work and is very user-friendly in that almost all communication with the program makes sense in terms of magnetics problems. TWODEPEP is much more mathematical; it is capable of solving a very broad variety of physical problems, not merely in magnetics or electromagnetics. It is not particularly well suited to magnetics, nor for that matter to any one physical application area; on the other hand, it is very flexible in the mathematical sense.

MAGGY

MAGGY is a mature software product with a long and distinguished history of in-house use within the Philips group of companies. As CAD software has come into wide use, programs such as MAGGY have gradually ceased to confer a unique competitive advantage in product development, and have instead become marketable products in their own right. MAGGY may well be a trend-setter, the first proprietary in-house magnetics program to be released in the CAD marketplace. It should be noted that MAGGY is not marketed by its developers, but by Infolytica Corporation, a specialist software house which provides user training and advisory services but refers internal maintenance problems to the program originators. In this respect, MAGGY resembles PE2D and FLUX2D above.

MAGGY is well suited to magnetics and has general functional characteristics comparable to those of the interactive programs discussed above. It uses quadrilateral finite elements to solve saturable-material static problems (which may include permanent magnets) in the x-y or r-z plane. Because it is a batch-run program, errors in input data are not immediately obvious through graphic screen echo—there is, after all, no graphics screen—so extensive internal data consistency checks are included in MAGGY. Output is available as numeric files, or pen-and-ink plotter tapes. Optionally, graphics screen devices may be substituted for the plotter for a more rapid examination of the results. MAGGY is unique in that parts of the system are furnished in compilable form, and it is possible for users to link in Fortran code of their own for any special purposes which may arise.

The computers on which MAGGY runs are relatively few, but they include the most popular ones: IBM and Amdahl mainframes, as well as Digital Equipment VAX-11 series machines. User licenses are normally issued for a three-year period. Almost all of the system is in the form of machine code, so that full maintenance is available from the vendors, in North America as well as Europe. License fees vary with the type of computer installation and are roughly comparable with the lower end of the interactive systems range.

Documentation supplied with MAGGY is relatively brief, but it is well written and contains sufficient information for users to experience little trouble. User training courses are available from the vendors.

TWODEPEP

TWODEPEP is a general-purpose finite element program capable of solving single or coupled partial differential equations, using a finite element method. High-order elements are used, and nonlinearities of the type encountered in magnetics can be dealt with. Input is by means of card images, and output is obtained from a line printer. It is also possible to have an output tape generated in a format suitable for driving a pen-and-ink plotter.

TWODEPEP is leased at an annual fee in the $2500 range and is delivered in source code form. Performance guarantees are naturally impossible to demand once a user has begun to make any modifications; on the other hand, the advisory service provided by the vendor (for installation assistance and advice) is reputed to be quite good.

The main strong point of TWODEPEP is its portability. Its input and output arrangements are not particularly well suited to any one type of problem, but it is capable of dealing with many physical problems, once they have been recast in purely mathematical form. It is written in Fortran and assumes nothing about locally available hardware; it can therefore be installed on a large number of different computers, under a wide variety of operating systems. Its user documentation is well written and not difficult to understand, with the proviso that TWODEPEP addresses itself strictly to partial differential equations of mathematical physics, not to computer-aided engineering design. In other words, it is likely to perform well for the analyst, not nearly so well for the designer.

The Literature of CAD in Magnetics

Design is an art and must be studied through examples like any other art. Engineering design naturally includes a large element of engineering science, which is best acquired in an analytic and deductive fashion. Expertise in the art of design, on the other hand, is best learned by case studies and by emulation of others. To aid the reader in finding examples related to specific classes of problems, this bibliography is provided.

Publications on Magnetics CAD

Computer-aided design first arrived in magnetics about 1970 and has generated an extensive published literature since that time. A comprehensive bibliography is not practical to offer within the compass of a book such as this one; and since the art is now rapidly advancing, it may in any case become obsolete very quickly. It appears more useful, therefore, to provide the reader a selective set of references, annotated to indicate the subject area and direction as well as the likely fields of application.

Professional Journals

A wide range of papers on CAD—containing examples of design procedure as well as reports of advances in mathematical methods—is published in professional journals. A considerable number, which may even amount to a majority, appears in the *IEEE Transactions on Magnetics*. Like most journals of this field, it deals with engineering applications of magnetics, including both the nature of magnetic materials and the ways in which they are used. The *IEE Proceedings* also carries numerous articles in this area, particularly in Parts A and B which are devoted to engineering science and electric power engineering, respectively. Fewer, but far from negligible numbers, of important contributions to the art also appear in the *Journal of Applied Physics*, and in the *Journal of Physics*, in its Part D (Applied Physics).

An early field of application for computer-aided design techniques in magnetics was the area of electric power equipment. While magnetic recording, control equipment design, robotics applications, geophysical electromagnetics, nondestructive testing, and dozens of other fields now account for a substantial fraction of the published literature, power equipment is probably still the most active single topic. Papers dealing with design in this area are usually published in the specialist journals, among which *IEEE Transactions on Power Apparatus and Systems* is a leading one; other significant ones include *Electric Machines and Electromechanics*.

Contrary to what might be expected, specialist journals such as the *International Journal for Numerical Methods in Engineering*, *Computer Methods in Applied Mechanics and Engineering*, and *Computer Aided Design* do not carry many articles of direct interest to people who actually wish to design things. They are much more strongly oriented to the underlying mathematical and computational techniques and are thus of greater value to software engineers specializing in computer-aided design.

Conferences

Much of the current material in any rapidly developing area is presented at professional conferences. Those relevant to magnetic device design, and international in scope, include Intermag, MMM, Compumag, and ICM.

Intermag is an annual event, sponsored by the IEEE Magnetics Society. It takes place in Europe, North America, and occasionally in Asia, on a rotation basis, usually in spring or early summer. The annual Conference on Magnetism and Magnetic Materials (MMM) is sponsored by the American Institute of Physics and takes place in summer or fall. Intermag and MMM are combined into a joint conference every three years. As might be expected, both conferences contain papers on material properties as well as analysis and design techniques; but MMM usually contains fewer papers on design than does Intermag.

Compumag is a separate conference, organized by an International Steering Committee which is not directly affiliated with any major professional organization. It exists specifically to deal with computational methods, and is therefore devoted to papers on new computational techniques, as well as papers on design methodology. Compumag takes place every two or three years, usually alternating between European and American locations.

ICM, the International Conference on Magnetism, is a large annual gathering somewhat more oriented to physics than engineering techniques. The aperiodic International Conference on Magnet Technology and the Applied Superconductivity Conferences are engineering-oriented, but somewhat specialized; the former deals principally with magnets used for high-energy physics research, the latter with superconductive magnets.

A frequent problem with conference papers in engineering is difficulty of access, for conference proceedings are often not available in libraries. The area of magnetic device design is fortunate in this respect, because the IEEE Magnetics Society had the foresight and vision in the 1970s to absorb the conference proceedings of Intermag, Compumag, and the Applied Superconductivity Conferences into its house journal, the *IEEE Transactions on Magnetics*. As a result, access to papers is easy and convenient. It is also for this reason that the *IEEE Transactions on Magnetics* appears to dominate publication in the area. A similar situation prevails with respect to MMM, whose papers are published in the *Journal of Applied Physics*.

There are many conferences and meetings devoted to application topics, in which papers on CAD appear in context with the purposes to which CAD systems have been put. Because CAD methods first became common in the power equipment industry and the nuclear physics community, conferences in those areas attract particularly substantial numbers of CAD-related papers. Of particular note are the Winter and Summer Power Meetings held by the IEEE Power Engineering Society. Papers presented at these meetings are automatically considered for the *IEEE Transactions on Power Apparatus and Systems*, and no separate volume of conference transactions is considered necessary.

Textbooks and Monographs

There are as yet no textbooks on computer-aided design of magnetic devices. There is only one textbook currently available in the English language on finite element analysis specifically aimed at electrical engineering. With growth and maturing of the field, such books will undoubtedly appear before long. For the moment, however, the designer must make do with periodical articles and conference papers.

The monograph literature of computer-aided design is quite large, but not of much use to the engineering designer. Almost all the available books in this area deal with methodology, that is, with how to design computer aided design systems, not how to design magnetic devices.

Two textbooks and monographs of relevance to computer-aided design in magnetics, and to the associated computational methods, are:

Silvester, P. P., Ferrari, R. L.: *Finite elements for electrical engineers.* ix + 209 pp. Cambridge: Cambridge University Press, 1983.

Chari, M. V. K., Silvester, P. P. [eds.]: *Finite elements in electric and magnetic field problems.* xii + 219 pp. Chichester: John Wiley & Sons, 1980.

An Annotated Bibliography

The bibliography given here is intended to serve as a guide, as a starting point for bibliographic research; it does not attempt to be exhaustive or complete. On the contrary, the authors have felt that the reader's needs would be better served by keeping the number of references comparatively small—but at the same time attaching an explanatory or indicative annotation to each. In some respects, this collection of references could well be regarded as an initial reading list, rather than as a bibliography.

Most of the items listed here serve to guide the reader in finding more detail in individual application areas, and to permit designers to exploit to the full the CAD systems now available. Some references, especially those dealing with the mathematical physics aspect of electromagnetics, transcend the present state of the art. A few others have been included because they confront issues still at the research forefront, and thereby serve to forecast what may be possible in tomorrow's systems. In keeping with these objectives, the list has been kept compact, omitting all references to journals not commonly held by engineering libraries, avoiding material in languages other than English, and eliminating all conference proceedings, collected volumes, and other comparatively ephemeral items.

The references are organized in chapter order, so they might well be regarded as notes for the individual chapters of this book. Within each chapter heading, they are simply listed alphabetically by the first author's surname.

Introduction

While there is now a very extensive and rapidly growing literature on the methods of CAD in magnetics, there is a surprising shortage of material on its industrial impact and on the way CAD is best used in manufacturing industries. The references given here include a substantial proportion of all the available literature.

Freeman, E. M.: Interactive computer aided design of electric machines and electromagnetic apparatus. *Journal of Applied Physics*, **53**, pp. 8393–8397, 1982.

The methods of computer aided design are reviewed; their implementation in large-scale computers and personal workstations is contrasted. The relationship of computer-aided design to computer-aided manufacturing is discussed briefly.

Lord, W.: Application of numerical field modelling to electromagnetic methods of nondestructive testing. *IEEE Transactions on Magnetics*, **MAG-19**, pp. 2437–2442, 1983.

The accepted techniques of nondestructive testing, particularly eddy current and magnetic remanence methods, are reviewed. Their implications to the construction of CAD software, to problem formulation, and to computational requirements, are explored briefly.

Okabe, M., Okada, M., Tsuchiya, H.: Effects of magnetic characteristics of materials in the iron loss in the three phase transformer core. *IEEE Transactions on Magnetics*, **MAG-19**, pp. 2192–2194, 1983.

This article illustrates how new design rules are derived from field analysis. The eddy current loss depends mainly on three numeric parameters: the Froehlich constant, the ratio of effective to maximum flux density, and the ratio of permeabilities in the principal directions of the core material.

Preston, T. W., Reece, A. B. J.: The contribution of the finite-element method to the design of electrical machines: an industrial viewpoint. *IEEE Transactions on Magnetics*, **MAG-19**, pp. 2375–2380, 1983.

An overview of the growth of finite element methods as well as of the requirements and the software used by the General Electric Company for design of large electric machines. Data preparation, postprocessing, and computer requirements are mentioned briefly.

Raby, K. F.: On getting the right answer. *IEE Proceedings*, **129**, pt. A, pp. 39–45, 1982.

Computer-aided design systems for magnetic device design are now becoming widespread in industry and should enter engineering education as well as practice at an early date. Examples of industrial design illustrate the argument.

Silvester, P. P.: Interactive computer aided design in magnetics. *IEEE Transactions on Magnetics*, **MAG-17**, pp. 3388–3392, 1981.

The computer-aided design process is seen as a formalized version of design processes of the pre-computer era. System structures, and the probable impact of new computer hardware, are reviewed, with particular reference to currently existing CAD software.

Magnetic Material Representation

The majority of commercially available CAD systems for magnetics presume single-valued magnetization curves and do not permit hysteresis to be modelled in detail. Permanent magnet materials may be included in analysis, however. It should be noted that data on the behavior of magnetically hard materials are often difficult to obtain for all directions except the direction parallel to the remanent magnetization; and when such data are available, they are often quite inaccurate.

Hysteresis models are of considerable interest to the magnetic recording community and, as the use of CAD systems in that community grows, improved hysteresis models are very likely to become available in CAD systems. For this reason, a few references on hysteretic behavior of materials are given both here and in the section below on problem formulation.

Baran, W.: Core magnet assembly with anisotropic Alnico permanent magnets. *IEEE Transactions on Magnetics*, **MAG-19**, pp. 2498–2501, 1983.

An extensive discussion of material modelling problems encountered with permanent magnet materials. Useful bibliography of material modelling.

Bhadra, S. N.: A direct method to predict instantaneous saturation curve from rms saturation curve. *IEEE Transactions on Magnetics*, **MAG-18**, pp.1867–1870, 1982.

Supposing $i = a_1\phi + a_5\phi^5$ to be a valid description of the mean magnetization curve of a transformer or similar device, the rms behavior of the device is analyzed. A method for finding the constants a_1 and a_5 from rms measurements is given.

Binns, K. J., Low, T. S., Jabbar, M. A.: Behaviour of a polymer-bonded rare-earth magnet under excitation in two directions at right angles. *IEE Proceedings*, **130**, pt. B, pp. 25–32, 1983.

B-H curves of the material are measured experimentally and embedded in a permeability tensor representation useful for finite element analysis. The method is verified by analysis and experiment.

Brauer, J. R., Larkin, L. A., Overbye, V. D.: Finite element modeling of permanent magnet devices. *Journal of Applied Physics*, **55**, pp. 2183–2185, 1984.

Permanent magnet materials are treated by substantially the same technique as given in this book: the demagnetization curve is shifted by an amount equal to the coercive force, to cross through the origin. The magnetization is then replaced by mathematically equivalent current densities. Both the x-y and r-z cases are briefly treated.

Campbell, P., Al-Murshid, S. A.: The effect of the magnetization distribution within anisotropic Alnico magnets upon field calculation. *IEEE Transactions on Magnetics*, **MAG-16**, 5, pp. 1032–1034, 1980.

In anisotropic magnetic materials, the remanence in the preferred direction is modelled by a distributed current density, while two *B-H* curves are used—one for the preferred direction and one orthogonal to it. Detailed data and measurements are given for a particular Alnico material.

Campbell, P., Al-Murshid, S. A.: A model of anisotropic Alnico magnets for field computation. *IEEE Transactions on Magnetics*, **MAG-18**, pp. 898–904, 1982.

An accurate, physically based tensor permeability model is derived and material data are furnished on Alnico 5, 5–7, and 8.

El-Sherbiny, M. K.: Representation of the magnetization characteristic by a sum of exponentials. *IEEE Transactions on Magnetics*, **MAG-9**, pp. 60–61, 1973.

$B = a_0 + a_1 \exp(-b_1 H) + \ldots + a_n \exp(-b_n H)$ is a form capable of modelling *B-H* curves well. The value $n = 4$ is suggested. Trial and error fitting is needed to determine the constants, and computational expense in subsequent use of this expression is great.

Forghani, B., Freeman, E. M., Lowther, D. A., Silvester, P. P.: Interactive modelling of magnetisation curves. *IEEE Transactions on Magnetics*, **MAG-18**, pp. 1070–1072, 1982.

The modelling subsystem within MagNet employs a six-piece Hermite interpolation polynomial fit. It is coupled to a full-fledged file management system for maintaining curve data.

Hanalla, A. Y., Macdonald, D. C.: Representation of soft magnetic materials. *IEE Proceedings*, **127**, pt. A, pp. 386–391, 1980.

A representation of hysteresis loops is given, which the authors believe to be simple enough for incorporation in finite element analysis.

Kamminga, W.: Finite-element solutions for devices with permanent magnets. *Journal of Physics D: Applied Physics*, **8**, pp. 841–855, 1975.

Permanent magnets are described in terms of the magnetization vector M and a preferred direction. The variational functional for axisymmetric cases is set up for an arbitrary preferred direction and discretized on first-order triangles (subject to the approximation that centroidal values of radius are used over entire triangles). Various examples are solved and checked against experiment; in all, the preferred direction is axial and the magnetization uniform.

Lakehal, M., Jufer, M.: Finite elements calculation of polarized, anisotropic and saturated electromechanical structures. *IEEE Transactions on Magnetics*, **MAG-19**, pp. 2491–2493, 1983.

A uniformly magnetized region, which could be one finite element, is shown to be equivalent to an unmagnetized region with surface currents flowing on it. This equivalence is used to construct finite element models for permanent magnets.

Lord, W.: Demagnetisation phenomena in high-performance d.c. servomotors. *IEE Proceedings*, **122**, pp. 392–394, 1975.

Discusses demagnetization curves of hard magnetic materials and their minor hysteresis loop behavior.

Sarma, M. S.: Analysis and design considerations for permanent-magnet aerospace-instrument devices. *IEEE Transactions on Magnetics*, **MAG-16**, pp. 797–799, 1980.

Samarium cobalt magnets are assumed to possess a coercive force H_c and to have an incremental permeability equal to that of free space. H_c is accommodated in finite element models by introducing an equivalent current-carrying coil.

Slomczynska, J.: Nonlinear analysis of the magnetic flux distribution in the magnetized magnet stabilized in air. *IEEE Transactions on Magnetics*, **MAG-10**, pp. 1109–1113, 1974.

Two-dimensional magnetization and subsequent stabilization of a magnet are treated assuming that the remanence has a fixed value $\mathbf{B_r}$, and that reluctivity at right angles to the remanence has a constant value equal to the recoil reluctivity along the direction of $\mathbf{B_r}$. In the direction of $\mathbf{B_r}$, reluctivity is assumed to follow the hysteresis loop.

Weiss, J., Garg, V. K., Shah, M., Sternheim, E.: Finite element analysis of magnetic fields with permanent magnet. *IEEE Transactions on Magnetics*, **MAG-20**, pp. 1933–1935, 1984.

A representation of hard magnetic materials as incorporated in the Westinghouse WEMAP system is described. It is substantially similar to that given in this book.

Widger, G. F. T.: Representation of magnetisation curves over extensive range by rational-fraction approximation. *IEE Proceedings*, **116**, pp. 156–160, 1969.

$B/H = (a_0 + a_1 + \ldots + a_n H^n)/(1 + \ldots + b_n H^n)$, or an analogous rational fraction expression for H/B in terms of B, usually reproduces B-H curves with more than adequate accuracy, with $n = 2$ or $n = 3$. An iterative process is used to find the coefficients a_i and b_i.

The Potential Equations of Magnetics

Potential theory is a well-established part of mathematical physics. Despite its ancient lineage, however, some potential formulations are better suited to numerical computation than others, and some are particularly desirable because the inherent nature of magnetics raises certain requirements quite different from, say, stress analysis or fluid flow. The references included here have been selected particularly with a view to their relevance in connection with existing software systems.

Burais, N., Foggia, A., Nicolas, A., Sabonnadiere, J. C.: Electromagnetic field formulation for eddy current calculations in nondestructive testing systems. *IEEE Transactions on Magnetics,* **MAG-18**, pp. 1058–1060, 1982.

For eddy current testing problems involving x-y plane currents, the z-directed component of **H** can usefully serve in lieu of a potential. Boundary conditions on longitudinal cracks are then of the Dirichlet type, and sensor impedances are readily calculated from the field solution.

Carpenter, C. J.: Numerical solution of magnetic fields in the vicinity of current-carrying conductors. *IEE Proceedings,* **114**, pp. 1793–1800, 1967.

The scalar potential method for magnetic fields can be extended to current-carrying regions, by defining **H** to consist of two components: one which has curl $\mathbf{H}_0 = \mathbf{J}$, but is otherwise arbitrary; and a second one which can be derived from a scalar potential. The method is applied in finite-difference form to several problems.

Chari, M. V. K., Csendes, Z. J.: Finite element analysis of the skin effect in current carrying conductors. *IEEE Transactions on Magnetics,* **MAG-13**, pp. 1125–1127, 1977.

The diffusion problem in linear media is set up in the usual form and restated as the condition of stationarity (the authors say minimality, without proof) of a quadratic functional on the complex vector potential **A**. Discretization on first-order triangles then yields the matrix equation earlier obtained by Galerkin methods. The requirements div $\mathbf{J} = 0$ and div $\mathbf{A} = 0$ are imposed by shifting the reference level of the potential.

Forghani, B., Lowther, D. A., Silvester, P. P., Stone, G. O.: Newton-Raphson finite element programs for axisymmetric vector fields. *IEEE Transactions on Magnetics,* **MAG-19**, pp. 2523–2526, 1983.

To keep finite element programs free of singularities, the quantity A_θ/r is used instead of the potential component A_θ. In nonlinear problems, the Jacobian matrices for the Newton process are then well behaved but require numerical integration because the permeability varies over an element.

Hammond, P.: Use of potentials in calculation of electromagnetic fields. *IEE Proceedings,* **129**, pt. A, pp. 106–112, 1982.

A detailed survey of the various potentials useful for magnetics and electromagnetics.

Lavers, J. D.: Finite element solution of nonlinear two dimensional TE-mode eddy current problems. *IEEE Transactions on Magnetics,* **MAG-19**, pp. 2201–2203, 1983.

The translationally uniform problem with eddy currents having only x-y plane components is stated using H_z as a potential function. The resulting finite element matrices are not symmetric, but yield to the biconjugate gradient solution method.

Mohammed, O. A., Demerdash, N. A., Nehl, T. W.: Validity of finite element formulation and solution of three dimensional magnetostatic problems in electrical devices with applications to transformers and reactors. *IEEE Transactions on Power Apparatus and Systems*, **PAS-103**, pp. 1846–1853, 1984.

For simply connected domains containing materials with positive definite reluctivity tensors, a unique vector potential formulation of the magnetic field is given. Uniqueness of the potential is ensured by imposing the correct boundary conditions at the edges of the domain.

Pillsbury, R. D., Jr.: A three dimensional eddy current formulation using two potentials: the magnetic vector potential and total magnetic scalar potential. *IEEE Transactions on Magnetics*, **MAG-19**, pp. 2284–2247, 1983.

The magnetic vector potential is used in all current-carrying portions of space, the scalar potential elsewhere. The two potentials are coupled by enforcing continuity of normal flux density and tangential magnetic field at all interface points.

Polak, S. J., Wachters, A. J. H., van Welij, J. S.: A new 3-D eddy current model. *IEEE Transactions on Magnetics*, **MAG-19**, pp. 2447–2449, 1983.

The problem space is partitioned into three regions. Where eddy currents flow, the magnetic vector potential **A** is used to represent the field; elsewhere, either the magnetic scalar potential or (in iron) the reduced scalar potential is used, to avoid numerical instability.

Renew, D. C., Blake, S., Jacobs, D. A. H., Rutter, P.: A magnetostatic 2 1/2-dimensional field calculation program for turbogenerators. *IEEE Transactions on Magnetics*, **MAG-19**, pp. 2317–2320, 1983.

A scalar potential formulation of the end-zone of an electric machine incorporates periodic angular variation of the potential and thereby permits separation of the true three-dimensional problem into a set of two-dimensional problems. The method can be given various numerical implementations.

Rodger, D, Eastham, J. F.: A formulation for low frequency eddy current solutions. *IEEE Transactions on Magnetics*, **MAG-19**, pp. 2443–2446, 1983.

The problem space is partitioned into regions which carry eddy currents and those which do not. Alternative formulations are given with either **A** or **H** as the fundamental variable in the first, and the magnetic scalar potential in the second.

Tokumasu, T., Doi, S., Ito, K., Yamamoto, M.: An electric vector potential method approach for 3-D electromagnetic field in turbine generator stator core end. *IEEE Transactions on Power Apparatus and Systems*, **PAS-103**, pp. 1330–1338, 1984.

A potential formulation dual to the usual ones is given, in which an electric vector potential and magnetic scalar potential are the fundamental quantities.

Trowbridge, C. W.: Three-dimensional field computation. *IEEE Transactions on Magnetics*, **MAG-18**, pp. 293–297, 1982.

A brief survey is given of the various potential formulations available, with a view to choosing those computationally efficient for three-dimensional problems.

Problem Modelling and Mesh Construction

The statement of problems in computational fashion and the construction of good-quality finite element meshes are questions of considerable importance not only to electromagnetics engineers, but to the structural analysis and fluid dynamics communities as well. While the differential equations underlying the several fields are relatively similar, the boundary conditions and side constraints which define the full problem are frequently of quite different character; furthermore, the needs of structural and magnetics engineers diverge considerably because they seek solutions for different purposes. Nevertheless, there is sufficient common ground for achievements in one field to be of interest to workers in the other. Accordingly, some key items of the structural analysis literature appear below, along with more strictly electromagnetic work.

Ancelle, B., Callagher, E., Masse, P.: ENTREE: a fully parametric preprocessor for computer aided design of magnetic devices. *IEEE Transactions on Magnetics,* **MAG-18**, pp. 630–632, 1982.

A description is given of the data structures and program structure of the ENTREE mesh preprocessor. ENTREE is not usable independently of FLUX, but forms part of that analysis system.

Berkery, J. F., Barton, R. K., Konrad, A.: Automatic finite element grid generation for motor design analysis. *IEEE Transactions on Magnetics,* **MAG-20**, pp. 1924–1926, 1984.

A mesh editing system is described, broadly similar to that of MagNet but incorporating particular software features that suit it to dc machine analysis.

Cendes, Z. J., Shenton, D., Shahnasser, H.: Magnetic field computation using Delaunay triangulation and complementary finite element methods. *IEEE Transactions on Magnetics,* **MAG-19**, pp. 2551–2554, 1983.

The Delaunay algorithm, which is now becoming fairly common in commercially available software, is explained in a very readable fashion.

Haber, R., Shephard, M. S., Abel, J. F., Gallagher, R. H., Greenberg, D. P.: A generalized graphic preprocessor for two-dimensional finite element analysis. *Computer Graphics,* **12**, pp. 323–329, 1978.

Accepts boundary and interface data; shreds interiors of regions using several schemes of coordinate interpolation. Allows subsequent editing of individual elements or nodes. As a separate module, an *attribute editor* then permits boundary conditions, loadings, and material properties to be attached to the geometric model.

Imafuku, I., Kodera, Y., Sayawaki, M., Kono, M.: A generalized automatic mesh generation scheme for the finite element method. *International Journal for Numerical Methods in Engineering,* **15**, pp. 713–732, 1980.

The large-scale finite element preprocessor of Mitsui Shipbuilding Ltd. accepts as input homogeneous regular blocks and prescriptions regarding how many nodes to place along each edge. The program then subdivides the union of all blocks into elements and prepares the output file, including node and element information. Algorithms used in the program are described in detail.

Lindholm, D. A.: Automatic triangular mesh generation on surfaces of polyhedra. *IEEE Transactions on Magnetics,* **MAG-19**, pp. 2539–2542, 1983.

A surface-triangulation program is described, which is useful for two-dimensional programs, or for three-dimensional ones as a starting point for automatic decomposition into tetrahedra.

Masse, Ph., Coulomb, J. L., Ancelle, B.: System design methodology in C.A.D. programs based on the finite element method. *IEEE Transactions on Magnetics,* **MAG-18**, pp. 609–616, 1982.

Methods and criteria underlying the structure of CAD systems are discussed, with particular attention to the data structures, input preprocessors, and structuring of preprocessor-solver interaction.

Molfino, P., Molinari, G., Viviani, A.: A user oriented modular package for the solution of general field problems under time varying conditions. *IEEE Transactions on Magnetics,* **MAG-18**, pp. 638–643, 1982.

COMPELL is a proprietary software system. It is batch-processing oriented but possesses a problem-oriented input language suited to magnetics. It includes an extensive mesh preprocessor and facilities for discretization and solution either by finite elements or finite differences.

Nelson, J. M.: A triangulation algorithm for arbitrary planar domains. *Applied Mathematical Modelling,* **2**, pp. 151–159, 1978.

Given the outline of a domain and a set of interior points, a triangulation is constructed. The algorithm is said to be faster than that of Suhara and Fukuda, because the criterion of triangle shape is such as to obviate most of the required checking for element overlaps.

Pammer, Z.: A mesh refinement method for transient heat conduction problems solved by finite elements. *International Journal for Numerical Methods in Engineering,* **15**, pp. 495–505, 1980.

Criteria are developed for choice of element size near the surface of a material subjected to a sudden change in ambient temperature.

Preis, K.: A contribution to eddy current calculations in plane and axisymmetric multiconductor systems. *IEEE Transactions on Magnetics,* **MAG-19**, pp. 2397–2400, 1983.

Finite element solutions are given for linear sinusoidal problems. Comparisons are made with integral finite element solutions, and some indications are obtained of the fineness of modelling required.

Shephard, M. S., Gallagher, R. H., Abel, J. F.: The synthesis of near-optimum finite element meshes with interactive computer graphics. *International Journal for Numerical Methods in Engineering,* **15**, pp. 1021–1039, 1980.

Automatic mesh refinement is based on approximate first derivatives of the energy density. Triangulation is automatic, with fixed edge nodes. Edge nodes are added to either automatically or manually, or both, as subdivision proceeds.

Sitaraman, V., Saxena, R. B.: Mesh grading in finite element analysis. *Electric Machines and Power Systems,* **8**, pp. 33–43, 1983.

An experimental algorithm for producing graded meshes in electric machine problems is described. Its behavior is illustrated by example meshes.

Thacker, W. C.: A brief review of techniques for generating irregular computational grids. *International Journal for Numerical Methods in Engineering,* **15**, pp. 1335–1341, 1980.

A concise but detailed review of the state of the art in mesh generation, three as well as two dimensional. The bibliography (80 items) is almost entirely composed of accessible journal literature, in English, but it is now somewhat dated.

Tracy, T. F.: Graphical pre- and post-processor for 2-dimensional finite element method programs. *Computer Graphics,* **11**, pp. 8–12, 1977.

The preprocessor accepts a shape description as a set of subregions, each defined by an edge sequence. Line segments may be inserted ad lib, the program keeps track of what constitutes a subregion. Automatic mesh generation—quadrilaterals and triangles—then follows. The method of problem definition is alluded to but not described. The postprocessor is essentially a straightforward, though good, display package.

Yildir, Y. B., Wexler, A.: MANDAP—a FEM/BEM data preparation package. *IEEE Transactions on Magnetics,* **MAG-19**, pp. 2562–2564, 1983.

A mesh generation package using the Coons patch methodology. The package is oriented to batch processing and does not appear to have an associated attribute editor.

Field Problems in Magnetic Devices

The art of formulating problems in a solvable way is central to every branch of the art of engineering design, and the formulation of field problems is no exception. Not surprisingly, references in this area cover a broad range of applications and are numerous.

Ashtiani, C. N., Lowther, D. A.: Simulation of the transient and subtransient reactances of a large hydrogenerator by finite elements. *IEEE Transactions on Power Apparatus and Systems,* **PAS-103**, pp. 1788–1794, 1984.

Because transient and subtransient time constants differ substantially, partial-regions solutions can be employed to find the fields, and hence the reactances. Results are compared with experimental measurements.

Barbero, V., Dal Mut, G., Grigoli, G., Santamaria, M.: Axisymmetric analysis and experimental measurements of magnetic field in the end region of a turbine generator. *IEEE Transactions on Magnetics,* **MAG-19**, pp. 2623–2627, 1983.

The generator is assumed to be axisymmetric and, despite the violation of physical laws implied thereby, the end-region fluxes are computed, using a vector potential formulation.

Barton, M. L.: Loss calculation in laminated steel utilizing anisotropic magnetic permeability. *IEEE Transactions on Power Apparatus and Systems,* **PAS-99**, pp. 1280–1287, 1980.

The anisotropic magnetic problem for linear materials is formulated in detail and solved, for example cases. Numerous plots are shown, and it is verified that an anisotropic representation well describes laminated iron.

Chiampi, M., Negro, A. L., Tartaglia, M.: Alternating electromagnetic field computation in laminated cores. *IEEE Transactions on Magnetics,* **MAG-19**, pp. 1530–1536, 1983.

Mitered gaps in laminated cores are examined to determine power loss due to eddy currents. The total flux and mmf are decomposed in Fourier series and the coupled nonlinear equations for all harmonics are solved simultaneously.

Demerdash, N. A., Nehl, T. W.: Use of numerical analysis of nonlinear eddy current problems by finite elements in the determination of parameters of electrical machines with solid iron rotors. *IEEE Transactions on Magnetics,* **MAG-15**, pp. 1482–1484, 1979.

Linearized equivalent permeability (based on rms values of *H* and *B*) is used in the finite element solution of alternator fields, so as to arrive at estimates of eddy currents and losses.

Demerdash, N. A., Nehl, T. W., Fouad, F. A., Arkadan, A. A.: Analysis of the magnetic field in rotating armature electrically commutated dc machines by finite elements. *IEEE Transactions on Power Apparatus and Systems,* **PAS-103**, pp. 1829–1836, 1984.

Equivalent circuit parameters, suitable for inclusion in circuit models for analyzing electronic commutation, are determined from field solutions.

Desserre, J. R., Marrocco, A.: Simulation of the writing process using finite element and augmented Lagrangian methods. *IEEE Transactions on Magnetics,* **MAG-18**, pp. 238–241, 1982.

Fields are computed in the neighborhood of a recording head and medium. Using a reasonably simple hysteresis model, the remanence in the recording medium is then found.

di Napoli, A., Paggi, R.: A model of anisotropic grain-oriented steel. *IEEE Transactions on Magnetics,* **MAG-19**, pp. 1557–1561, 1983.

Material properties in the principal directions suffice to define a model which yields *B-H* curves of adequate accuracy in any direction of magnetization.

Foggia, A., Sabonnadière, J. C., Silvester, P.: Finite element solution of saturated travelling magnetic field problems. *IEEE Transactions on Power Apparatus and Systems,* **PAS-94**, pp. 866–871, 1975.

The linear induction machine problem, involving induced voltage and motional (convection) terms, is discretized using the Galerkin approach. First-order triangular elements in conjunction with a time-stepping transient integration are used for solution.

Hammond, P., Tsiboukis, T. D.: Dual finite-element calculations for static electric and magnetic fields. *IEE Proceedings,* **130**, pt. A, pp. 105–111, 1983.

Dual solutions on the same finite element mesh give upper and lower bounds for stored energy, hence for inductances, capacitances, and related quantities.

Hanalla, A. Y., Macdonald, D. C.: A nodal method for the numerical solution of transient field problems in electrical machines. *IEEE Transactions on Magnetics,* **MAG-11**, pp. 1544–1546, 1975.

A variant first-order triangular element method is described for the scalar diffusion equation in **A**. Two concurrent approximations are used for **A**: the standard linear interpolation for the Laplacian, and a "nodal attribution" (similar to mass lumping in structural engineering) for the time-dependent term and for the current densities. The results are said to be better than for self-consistent methods, although no evidence is presented.

Hanalla, A. Y., Macdonald, D. C.: Numerical analysis of transient field problems in electrical machines. *IEE Proceedings,* **123**, pp. 893–898, 1976.

The two-dimensional saturable magnetic field, with eddy currents included, yields the diffusion equation in **A**. Discretization by first-order triangles is used for the spatial solution at each time step. Discretization of the diffusion term follows the same unconventional nodal lumping technique as used by the same authors earlier for the impressed current density.

Koehler, T. R.: Self-consistent vector calculation of magnetic recording using the finite element method. *Journal of Applied Physics,* **55**, pp. 2214–2216, 1984.

A conventional finite element representation of the recording head is used, with a hysteresis model in the recording medium. The remanence is then included by the introduction of equivalent sources in each element.

Konrad, A.: Integrodifferential finite element formulation of two-dimensional steady-state skin effect problems. *IEEE Transactions on Magnetics,* **MAG-18**, pp. 284–292, 1982.

The classical skin-effect problem is reformulated to employ the conductor total current as the forcing function. An extensive bibliography is given.

Konrad, A., Silvester, P. P.: Scalar finite element solution of magnetic fields in axisymmetric boundaries. *IEEE Transactions on Magnetics,* **MAG-18**, pp. 270–274, 1982.

If magnetic materials are assumed linear, any azimuthal variation in source densities can be decomposed in a Fourier series so that the three-dimensional boundary-value problem with axisymmetric boundaries is solvable as a set of separate two-dimensional problems, one for each Fourier series term.

Koshimoto, Y., Mikazuki, T.: High output reproduction characteristics of film heads. *IEEE Transactions on Magnetics,* **MAG-18**, pp. 1134–1136, 1982.

A Froelich-type approximation to the head material properties and a detailed consideration of recording medium remanence lead to a full finite element analysis of a two-dimensional head in terms of a magnetic scalar potential.

Lefranc, C., Feliachi, M., Coulomb, J. L.: On numerical modelling of hysteresis in recording systems. *IEEE Transactions on Magnetics,* **MAG-20**, pp. 1891–1893, 1984.

Hysteretic behavior is included in a finite element model by introducing appropriate coercive forces into an anisotropic material model. The recording process is simulated, and the readback signals of a particular recording system are computed.

Luetke-Daldrup, B.: Comparison of exact and approximate finite-element solution of the two-dimensional nonlinear eddy-current problem with measurements. *IEEE Transactions on Magnetics,* **MAG-20**, pp. 1936–1938, 1984.

Skin effect of a sinusoidal current in a slot conductor is solved taking iron saturation into account, first by time-stepping through the ac cycle and computing the instantaneous values, then by taking each finite element to have a constant permeability equal to that obtained from the mean magnetization curve for a flux density equal to three-quarters of the peak value. The two calculations are in excellent agreement with each other and experiments, suggesting that the second, simpler, method should be routinely used.

MacBain, J. A.: Magnetic field simulation from a voltage source. *IEEE Transactions on Magnetics*, **MAG-19**, pp. 2180–2182, 1983.

The generated (internal) voltage of a coil is computed from the field solution on a flux linkage basis. A proprietary program, not extensively described, is used.

Minnich, S. H., Tandon, S. C., Atkinson, D. R.: Comparison of two methods for modeling large-signal alternating magnetic fields using finite-elements. *IEEE Transactions on Power Apparatus and Systems*, **PAS-103**, pp. 2953–2960, 1984.

Explicit time integration is used to solve diffusion (eddy current) problems in steady state. This method is compared with results obtained by assuming material linearity, with each element having an effective permeability equal to that obtained as the ratio of peak B to peak H. The latter method shows considerable promise, but the definition of effective permeability requires further investigation.

Naito, T., Okada, T., Naito, T., Moriyasu, S.: Analysis of damper winding current of synchronous generator due to space subharmonic M.M.F. *IEEE Transactions on Magnetics*, **MAG-19**, pp. 2643–2646, 1983.

The problem of subharmonic mmf's is reformulated to permit modelling of only one pole pitch of a machine in order to determine fields.

Nakata, T., Takahashi, N., Kawase, Y., Nakano, M.: Influence of lamination orientation and stacking on magnetic characteristics of grain-oriented silicon steel laminations. *IEEE Transactions on Magnetics*, **MAG-20**, pp. 1774–1776, 1984.

Transformer cores containing butt joints in which anisotropic (grain-oriented) laminations are stacked in alternate directions are replaced by a fictitious homogeneous material, so as to allow two-dimensional analysis of the flux distribution in the core.

Nicolas, A., Sabonnadière, J. C., Silvester, P. P.: Three-dimensional solution of high-speed magnetic levitation problems. *Journal of Applied Physics*, **53**, pp. 8417–8419, 1982.

Because all magnetic material boundaries are parallel to the z axis, the source distributions and fields can be decomposed in orthogonal function series in the z direction, thereby decomposing the three-dimensional problem into a set of independent two-dimensional problems in the x-y plane. These are solved using conventional finite element techniques.

Nomura, T., Nakamura, S., Uda, S., Iwamoto, M., Okada, M.: Finite element analysis of operational impedance of superconducting alternators. *IEEE Transactions on Magnetics*, **MAG-19**, pp. 2608–2610, 1983.

Magnetic fields in the machine cross-section are described by a scalar potential, while the eddy currents in rotor damper bars are described by a stream function. Representations adequate for stability analysis are obtained.

Okuda, H., Abukawa, T., Arai, K., Ito, M.: Characteristics of ring permanent magnet bearing. *IEEE Transactions on Magnetics*, **MAG-20**, pp. 1693–1695, 1984.

The performance of a magnetic bearing for slightly eccentric placement is computed by constructing three axisymmetric finite element models and employing a force computation scheme which interpolates between the solutions. The true three-dimensional problem is therewith reduced to two dimensions.

Palmer, D. C., McDaniel, T. W.: Spurious signal pickup from the outer rails of a ferrite recording head. *IEEE Transactions on Magnetics*, **MAG-20**, pp. 912–914, 1984.

Two- and three-dimensional analyses of a Winchester-style monolithic head are compared to determine the validity of a two-dimensional solution.

Pearson, R. F., Knowles, J. E.: Application of MAGGY2 to short wavelength recording. *IEEE Transactions on Magnetics*, **MAG-19**, pp. 3456–3458, 1983.

Symmetry boundary conditions are applied to the recording problem to ensure that only one wavelength of the recorded tape need be considered. The recording head is assumed to be made of magnetically linear material.

Potter, P. G., Cambrell, G. K.: A combined finite element and loop analysis for nonlinearly interacting magnetic fields and circuits. *IEEE Transactions on Magnetics*, **MAG-19**, pp. 2352–2355, 1983.

Circuit equations for the terminal connections of a magnetic device augment the finite element equations for the field, as constraints introduced into the finite element model for transient solutions.

Reece, A. B. J., Khan, G. K. M., Chant, M. J.: Generator parameter prediction: finite-element simulation of the variable-frequency injection test. *GEC Journal of Science and Technology*, **49**, pp. 13–17, 1983.

The test simulation is described and its results compared with measurements, for machines of 380 MVA and 600 MVA rating. Some difficulties arise in representing contact resistance and end effects, but adequate corrections can be made to allow two-dimensional analyses to be used.

Sarma, M. S., Stahl, P., Ward, A.: Magnetic-field analysis of ferrofluidic seals for optimum design. *Journal of Applied Physics*, **55**, pp. 2595–2597, 1984.

The flux distribution in and around the ferrofluid is found by conventional axisymmetric finite element analysis. The rare-earth material that forms magnetic part of the bearing is modelled by giving it a constant permeability and a known remanence.

Sawada, T., Sakuma, T., Yoneda, K.: Yoke type MR head on perpendicular double layer media. *IEEE Transactions on Magnetics*, **MAG-20**, pp. 857–859, 1984.

A two-dimensional finite element analysis is augmented by giving the recording medium a square-wave magnetization. Recording performance characteristics are derived thereby.

Shinagawa, K., Fujiwara, H., Kugiya, F., Okuwaki, T., Kudo, M.: Simulation of perpendicular recording on Co-Cr media with a thin Permalloy film-ferrite composite head. *Journal of Applied Physics*, **55**, pp. 2585–2587, 1984.

The writing process is modelled by conventional finite element analysis, with the remanence in the recording medium modelled using a magnetization vector **M**. Many numerical experiments show that fully detailed modelling of the head shape is necessary, but that the permeability variation in the head material is small enough to warrant simplifying the analysis, by assigning constant permeability to the head.

Silvester, P., Haslam, C. R. S.: Magnetotelluric modelling by the finite element method. *Geophysical Prospecting,* **20**, pp. 872–891, 1972.

The linear two-dimensional complex diffusion equation is discretized on high-order elements using Galerkin projections. Surface potentials measurable in geophysics are related to directional derivatives of the magnetic vector potential. Directional differentiation is shown to be possible by giving the differentiation operation a matrix representation. The necessary geometry-independent matrices are tabulated.

Silvester, P., Konrad, A., Coulomb, J. L., Sabonnadiere, J. C.: Modal network representation of slot-embedded conductors. *Electric Machines and Electromechanics,* **1**, pp. 107–122, 1977.

A slot conductor can be described by a Foster-form *RL* network, in which each *RL* branch corresponds to one of the eigenfunctions of a boundary-value problem, in which the region for the eigenproblem is the slot cross-section. The method for finding the *R* and *L* values is given and compared with experimental results.

Silvester, P. P., Lowther, D. A., Carpenter, C. J., Wyatt, E. A.: Exterior finite elements for 2-dimensional field problems. *IEE Proceedings,* **124**, pp. 1267–1270, 1977.

A recursion technique is defined for potential problems, by which large exterior regions are modelled. The method requires defining an annular region outside the region of immediate interest and "balloons" this region to cover a very large region of the exterior space.

Simkin, J., Trowbridge, C. W.: On the use of the total scalar potential in the numerical solution of field problems in electromagnetics. *International Journal for Numerical Methods in Engineering,* **14**, pp. 423–440, 1979.

The various potential formulations of magnetostatics are reviewed. In problems involving both current-carrying nonferrous regions and iron parts, it is advantageous to use a reduced scalar potential in the former, total scalar potential in the latter. The two potentials are related at the region interface by a potential jump which depends on the source field H_s. The formulation given is valid for three dimensions, but results are reported for a two-dimensional test program.

Tandon, S. C.: Finite element analysis of induction machines. *IEEE Transactions on Magnetics,* **MAG-18**, pp.1722–1724, 1982.

Equivalent circuit parameters are obtained from static magnetic field analyses which simulate the synchronous running (no-load) test and the locked-rotor test of a machine. Using the equivalent circuit, loaded performance is then predicted accurately.

Tandon, S. C., Armor, A. F., Chari, M. V. K.: Nonlinear transient finite element field computation for electrical machines and devices. *IEEE Transactions on Power Apparatus and Systems,* **PAS-102**, pp. 1089–1095, 1983.

Time-stepping methods (implicit forward difference and Crank-Nicolson) are investigated for time-dependent formulations of saturable-material problems.

Turner, P. J., Macdonald, D. C.: Finite element prediction of steady state operation of a turbine-generator using conductivity matrices. *IEEE Transactions on Magnetics,* **MAG-19**, pp. 3262–3264, 1983.

Current densities in the cross-sectional representation of a machine are combined into nodal variables in a manner analogous to "mass matrix lumping" in structural analysis. Curiously, no inaccuracy appears to arise from this approximation, and test results correlate well with predictions on a 500 MVA machine.

Williamson, S., Ralph, J. W.: Finite-element analysis of an induction motor fed from a constant-voltage source. *IEE Proceedings,* **130**, pt. B, pp. 18–24, 1983.

Rotor slot skewing and iron loss are included in the derivation of equivalent circuit models.

Williamson, S. J., Ralph, J. W.: Finite element analysis for nonlinear magnetic field problems with complex current sources. *IEE Proceedings,* **129**, pt. A, pp. 391–395, 1982.

The sinusoidal steady state is treated by ignoring harmonics. The real and imaginary parts of current are taken at their rms values and treated as separate time-invariant quantities, coupled through the nonlinear magnetic properties of iron. No experimental assessment of the accuracy of this method is given.

Yamada, S., Kanamaru, Y., Bessho, K.: The transient magnetization process and operations in the plunger type electromagnet. *IEEE Transactions on Magnetics,* **MAG-12**, pp. 1056–1058, 1976.

The flux density distribution and its variation with time in an axisymmetric actuator are determined by finite element analysis. Experimental (search coil) results are given by way of comparison.

Yamada, H., Fujibayashi, K., Ito, K., Nakano, M., Kanai, H.: Calculation of loss in solid-salient-pole machine. *IEEE Transactions on Power Apparatus and Systems,* **PAS-103**, pp. 1355–1362, 1984.

Steady-state sinusoidal analysis uses a magnetostatic solution carried out simultaneously for real and imaginary parts, the two being coupled through the material reluctivity. The permeability in each element is taken as equal to that for the prevailing peak flux density. Eddy current losses are then estimated.

Postprocessing Operations in CAD

Since postprocessing is the art of extracting useful information from field solutions, postprocessing is the area of CAD of main interest to designers. Work in postprocessing is often difficult to separate from work in problem formulation, for clever postprocessing techniques often presuppose that the problem is set up in a particular way in the first place. Consequently, the references given in connection with these two topics overlap to a significant extent.

Ashtiani, C. N., Lowther, D. A.: The use of finite elements in the simulation of the steady state operation of a synchronous generator with a known terminal loading condition. *IEEE Transactions on Magnetics,* **MAG-19**, pp. 2381–2384, 1983.

A method is given for finding load angle and field current for given terminal conditions of salient-pole machines.

Ashtiani, C. N., Lowther, D. A.: Simulation of the steady-state reactances of a large water-wheel generator by finite elements. *IEEE Transactions on Power Apparatus and Systems,* **PAS-103**, pp. 1781–1787, 1984.

The steady-state reactances of a machine are computed on a flux linkage basis. Results are compared with experimental measurements on a large waterwheel generator.

Baden Fuller, A. J., dos Santos, M. L. X.: New method for the display of three-dimensional vector fields. *IEE Proceedings,* **127**, pt. A, pp. 435–442, 1980.

A rectangular pyramidal arrow is used to display the vector. For full EM field display two arrows are used, one for **E**, the other for **H**.

Bhargava, S. C.: Negative-sequence currents, losses, and temperature rise in the rotor of a turbogenerator during transient unbalanced operation. *Electric Machines and Power Systems,* **8**, pp. 155–168, 1983.

A two-dimensional finite element analysis yields fields from which the sequence currents are calculated and the various other quantities subsequently derived.

Biedinger, M.-M., Kant, M.: On the optimization of magnetic field sources in electromechanical energy conversion. *Journal of Applied Physics,* **53**, pp. 7061–7070, 1982.

Assuming magnetic materials to be linear, optimization can be carried out by solving a set of partial problems and subsequently superposing their solutions.

Campbell, P., Chari, M. V. K., D'Angelo, J.: Three-dimensional finite element solution of permanent magnet machines. *IEEE Transactions on Magnetics,* **MAG-17**, pp. 2997–2999, 1983.

Fields in permanent magnet machines are illustrated by plotting scalar equipotentials and by giving surface vector plots using conical Nassif arrowheads.

Coulomb, J. L.: A methodology for the determination of global electromechanical quantities from a finite element analysis and its application to the evaluation of magnetic forces, torques and stiffness. *IEEE Transactions on Magnetics,* **MAG-19**, pp. 2514–2519, 1983.

Two related methods are given, both based on the virtual work principle. A very simple example is used by way of illustration.

Coulomb, J. L., Meunier, G.: Finite element implementation of virtual work principle for magnetic or electric force and torque computation. *IEEE Transactions on Magnetics,* **MAG-20**, pp. 1894–1896, 1984.

The virtual work technique of force calculation is formalized within the finite element framework, and methods for computing forces are given in detail.

Dawson, G. E., Eastham, A. R., Ong, R.: Computer-aided design studies of the homopolar linear synchronous motor. *IEEE Transactions on Magnetics,* **MAG-20**, pp. 1927–1929, 1984.

Field analysis of the linear machine is followed by computations to determine the d-axis and q-axis reactances.

Demerdash, N. A., Fouad, F. A., Nehl, T. W.: Determination of winding inductances in ferrite type permanent magnet electric machinery by finite elements. *IEEE Transactions on Magnetics,* **MAG-18**, pp. 1052–1054, 1982.

The definitions of apparent (total) and incremental inductance are reviewed, and a technique for calculating their values is presented. Results for a small permanent magnet machine compare well with measurements.

Di Napoli, A.: Induction machine equivalent network parameters computation from electrical and magnetic fields analysis. *IEEE Transactions on Magnetics,* **MAG-15**, pp. 1470–1472, 1979.

The nonlinear vector potential equation (including speed term) of an induction machine is discretized. The discrete form is related to terminal quantities and hence to equivalent circuit parameters. These are then used to predict the speed-torque curve, which is checked experimentally.

Dougherty, J. W., Minnich, S. H.: Operational inductances of turbine generators, test data versus finite-element calculations. *IEEE Transactions on Power Apparatus and Systems,* **PAS-102**, pp. 3393–3404, 1983.

Incremental inductances around various operating points are computed on the basis of local incremental permeabilities and are compared with extensive tests on a 555 MVA generator. Detailed results are given.

Eastham, J. F., Rodger, D.: The performance of induction levitators. *IEEE Transactions on Magnetics,* **MAG-20**, pp. 1684–1686, 1984.

Finite element force calculations for an induced-current levitation magnet are compared with experimental results.

Flores, J. C., Buckley, G. W., McPherson, G., Jr.: The effects of saturation on the armature leakage reactance of large synchronous machines. *IEEE Transactions on Power Apparatus and Systems,* **PAS-103**, pp. 593–600, 1984.

Equations for the various machine reactances, in terms of field quantities obtainable from finite element solutions, are set up and specialized to the case of two-dimensional fields.

Helszajn, J., Baars, R. D., Nisbet, W. T.: Characteristics of circulators using planar triangular and disk resonators loaded with magnetic ridges. *IEEE Transactions on Microwave Theory and Techniques,* **MTT-28**, pp. 616–621, 1980.

Standard fourth-order triangular elements are used—the program appears to be that of Csendes and Silvester. The idea of an "eigennetwork" (the planar network defined by null lines) is introduced as a computable and useful concept.

Howe, D., Low, W. F.: Design and dynamic calculations for miniature permanent magnet stepper motors. *IEEE Transactions on Magnetics,* **MAG-20**, pp. 1768–1770, 1984.

Force calculations using Maxwell stresses are sensitive to mesh detail. Mesh artifact is reduced or eliminated by taking pairs of integration paths with opposite asymmetry characteristics and averaging the results.

Jacobus, A., Muller, W.: Numerical solution of forces and torques. *IEEE Transactions on Magnetics,* **MAG-19**, pp. 2589–2592, 1983.

Methods for computing forces from field solutions are briefly reviewed and recommendations made.

Kincaid, T. G., Chari, M. V. K.: The application of finite element method analysis to eddy current nondestructive evaluation. *IEEE Transactions on Magnetics,* **MAG-15**, pp. 1956–1960, 1979.

First-order finite elements are used to solve the complex (sinusoidal steady-state) eddy current diffusion problem near a long crack in a two-dimensional block. Details are given for calculating the coil terminal impedance from the computed fields.

Konrad, A.: Linear accelerator cavity field calculation by the finite element method. *IEEE Transactions on Nuclear Science,* **NS-20**, pp. 802–808, 1973.

A program, and the postprocessing techniques used in it, are described, for calculating the axisymmetric modes in axisymmetric resonant cavities which may possess a drift tube. Triangular elements of high order are employed.

Lowther, D., McFee, S., Sabourin, R.: Some aspects of the computer aided design of recording heads. *IEEE Transactions on Magnetics,* **MAG-20**, pp. 1980–1982, 1984.

Superposition of solutions and a variety of postprocessing computations can be carried out in recording head design with a commercially available postprocessing system. Various displays available from the system are shown.

Minnich, S. H., Chari, M. V. K., Berkery, J. F.: Operational impedances of turbine-generators by the finite-element method. *IEEE Transactions on Power Apparatus and Systems,* **PAS-102**, pp. 20–27, 1983.

Small-signal operational inductances (incremental inductances) are defined and computed in terms of the machine incremental permeabilities at the operating point.

Naito, T., Matsuki, J., Okada, T., Uenosono, C.: Analysis of damper winding currents in salient-pole generator taking into account the magnetic saturation. *Electrical Engineering in Japan,* **102**, pp. 107–114, 1980.

Damper bar inductances are calculated from the field solution. Loop equations are then written to relate the currents in the bars, and solved to find the bar currents.

Nakata, T.: Numerical analysis of flux and loss distributions in electrical machinery. *IEEE Transactions on Magnetics,* **MAG-20**, pp. 1750–1755, 1984.

A review of the author's work in finite element analysis of transformers, particularly with regard to anisotropic materials and methods of loss estimation.

Nassif, N., Silvester, P. P.: Graphic representations of three-component vector fields. *Computer-Aided Design,* **12**, pp. 289–294, 1980.

In a display plane, a vector is represented by means of an "arrowhead", an axisymmetric object with a unique representation from every viewing angle. Short cylinders with one face black are proposed as easy to understand and economic of screen space.

Palanisamy, R., Lord, W.: Prediction of eddy current signals for nondestructive testing of condenser tubing. *IEEE Transactions on Magnetics,* **MAG-19**, pp. 2213–2215, 1983.

An approximate complex finite element solution for an eddy current problem yields simulated test probe signals which can be graphically displayed to mimic oscilloscope traces of their experimental counterparts. Flaw detection is entirely dependent on the right form of graphic display being included in the postprocessor, for it requires a human operator to recognize a characteristic pattern.

Palka, R.: Application of the finite element technique to continuation problems of stationary fields. *IEEE Transactions on Magnetics,* **MAG-19**, pp. 2356–2359, 1983.

Experimentally or computationally derived results in part of a region can be used as boundary conditions to extrapolate the potential values. In general, an inverse problem results. Stability and accuracy of finite element solutions to inverse problems are discussed in this context.

Penman, J., Fraser, J. R., Smith, J. R., Grieve, M. D.: Complementary energy methods in the computation of electrostatic fields. *IEEE Transactions on Magnetics,* **MAG-19**, pp. 2288–2291, 1983.

Interesting illustrative use of arrows for the display of planar **E** and **D** fields. Otherwise, an extension of the authors' previous work on magnetostatics into the electrostatics area.

Piriou, F., Razek, A.: Calculation of saturated inductances for numerical simulation of synchronous machines. *IEEE Transactions on Magnetics,* **MAG-19**, pp. 2628–2631, 1983.

The nonsinusoidal behavior of machine fluxes is determined by a stepwise time integration, and inductances are computed from a flux-linkage-based definition.

Rafinejad, P., Sabonnadiere, J. C.: Finite element computer programs in design of electromagnetic devices. *IEEE Transactions on Magnetics,* **MAG-12**, pp. 575–578, 1976.

Electromechanical forces may be formulated in terms of dS/dx, where S is the Dirichlet matrix and x the direction of a virtual displacement. Reactances may be found from flux linkages. A contactor magnet is analyzed by way of example.

Rajanathan, C. B., Lowther, D. A., Freeman, E. M.: Finite-element analysis of the Xi-core levitator. *IEE Proceedings,* **131**, pp. 62–66, 1984.

Stress distributions in the levitated plate are computed and displayed in an interesting fashion by means of vector arrows. The finite element field analysis follows fairly conventional lines.

Silvester, P., Konrad, A.: Analysis of transformer leakage phenomena by high-order finite elements. *IEEE Transactions on Power Apparatus and Systems,* **PAS-92**, pp. 1843–1855, 1973.

Axisymmetric infinite-permeability transformers are analyzed by means of high-order finite elements. The leakage inductance, forces or coils, and other quantities of interest to designers can be expressed in terms of certain universal matrices; numeric values of these are tabulated up to fifth order.

Szczech, T. J., Perry, D. M., Palmquist, K. E.: Improved field equations for ring heads. *IEEE Transactions on Magnetics,* **MAG-19**, pp. 1740–1744, 1983.

Individual Cartesian components of flux density or field are of interest and value in magnetic recording. Several forms of graphic display of flux density and field vectors in and around the recording medium are employed.

Strangas, E. M., Hamilton, H. B.: A model for the chopper controlled dc series motor. *IEEE Transactions on Power Apparatus and Systems,* **PAS-102**, pp. 1403–1407, 1983.

Fields in the motor are computed for various time instants. The field results are used to find terminal parameters, and torque is determined by means of the Maxwell stress tensor.

Takeda, Y., Yagisawa, T., Suyama, A., Yamamoto, M.: Application of magnetic wedges to large motors. *IEEE Transactions on Magnetics,* **MAG-20**, pp. 1780–1782, 1984.

A finite element calculation is used to find the effective permeability of a slotted rotor with magnetic slot wedges, the ripple flux associated with it, and the resulting pole surface losses.

Yamada, H., Fujibayashi, K.: Torque of solid salient pole machine by finite element method. *Journal of Applied Physics*, **53**, pp. 8375–8377, 1982.

Torque is found by evaluation of the Poynting vector so as to determine the rotor equivalent impedance.

Weiss, J., Garg, V. K., Sternheim, E.: Eddy current loss calculation in multiconductor systems. *IEEE Transactions on Magnetics,* **MAG-19**, pp. 2207–2209, 1983.

Power is calculated directly from the computed current density, and forces are found by evaluating the product $\mathbf{J} \times \mathbf{B}$. Conductor impedance is unfortunately only formulated for a single conductor.

Postprocessing Systems for Magnetics

Much of the literature of postprocessing systems is embedded in general descriptive literature on CAD systems, a bit here and a bit there. The references cited here represent a substantial fraction of the existing literature.

Barton, R. K., Konrad, A., Berkery, J. F.: Finite element postprocessing for direct current machines. *IEEE Transactions on Magnetics,* **MAG-20**, pp. 1798–1800, 1984.

A small special-purpose postprocessing system is described, which computes a number of quantities of interest to the dc machine designer and which is capable of various graphic representations (e.g., flux plots).

Lowther, D. A., Silvester, P. P., Freeman, E. M.: The use of interactive postprocessing in the design of electromagnetic devices. *IEEE Transactions on Magnetics,* **MAG-16**, pp. 803–805, 1980.

An outline description of an early MagNet postprocessor, as implemented on a single-user workstation with fast interactive graphics support. Various graphs and plots of the solution may be placed on the screen, and arithmetic operations may be performed on the solution.

Lowther, D. A., Silvester, P. P., Freeman, E. M., Rea, K., Trowbridge,C. W., Newman, M., Simkin, J.: Ruthless—a general purpose finite element postprocessor. *IEEE Transactions on Magnetics,* **MAG-17**, pp. 3402–3404, 1981.

Functional description of a postprocessing system intended for use with the PE2D finite element software system, or with other software employing triangular finite elements.

Lowther, D., Rea, K., Silvester, P., Freeman, E., Trowbridge, C., Simkin, J., Newman, M.: A stack configured vector calculator for electromagnetic field evaluation. *IEEE Transactions on Magnetics,* **MAG-18**, pp. 627–629, 1982.

Design problems of stack-configured postprocessing subsystems are discussed, with particular reference to the user interface. Possible solutions are illustrated by examples from the Ruthless postprocessor.

Lowther, D. A., Forghani, B.: Interactive postprocessing techniques for electromagnetic field analysis. *IEEE Transactions on Magnetics,* **MAG-19**, pp. 2168–2170, 1983.

A stack-configured graphics-oriented interactive postprocessing system is described, and the methods for programming it are stated. The history of systems of this type is reviewed briefly. A few examples of programming and graphic displays are shown.

CAD Systems Present and Future

In addition to the commercial software systems now available for magnetics CAD, several other categories of programs are described in the literature. First, various major manufacturing organizations maintain proprietary programs for in-house use only. Although not available to outsiders, their authors are often keen to have the existence of these systems widely known. Secondly, university groups and workers at publicly supported research laboratories have made a quantity of software readily available, either through the established libraries or depositories, or by contacting the authors directly. Such programs are of course unmaintained, and usually either specialized to a particular application or subject to the peculiar requirements of an educational setting. Finally, some large-scale commercial software systems intended for civil engineering use have in recent years been equipped with add-on extensions for magnetics use. Where appropriate in the references below, an indication has been appended to the annotation to indicate the commercial status or nature of the software.

Commercial software in general lags behind proprietary software, because commercial software must be capable of performing in a broad variety of applications and in inexpert hands, with minimum attention, while the proprietary software user in a large organization often has easy access to the program originator. Not infrequently, software matures and grows more robust with use. Even though proprietary software may not be available to outsiders, a knowledge of what exists in the proprietary sphere may be useful to forecast the commercial software of tomorrow.

Andersen, O. W.: Transformer leakage flux program based on finite element method. *IEEE Transactions on Power Apparatus and Systems,* **PAS-92**, pp. 682–689, 1973.

First-order axisymmetric finite elements are used for the air and copper regions; iron is assumed infinitely permeable. The mesh is of fixed topology (rectangular); its geometry is variable. (Commercial software)

Armstrong, A. G. A. M., Biddlecombe, C. S.: The PE2D package for transient eddy current analysis. *IEEE Transactions on Magnetics,* **MAG-18**, pp. 411–415, 1982.

An outline description, with illustrative problems, of the main aspects of the PE2D package, particularly including time dependence and nonlinearity. (Commercial software)

Biddlecombe, C. S., Edwards, C. B., Shaw, M. J.: Use of the computer code PE2D in the electrostatic modelling of an electron-beam-generator/vacuum-diode interface. *IEE Proceedings,* **129**, pt. A, pp. 337–339, 1982.

The electric scalar potential is used to perform the modelling, and design is carried out through examination of successive trial equipotential plots. (Commercial software)

Biddlecombe, C. S., Simkin, J.: Enhancements to the PE2D package. *IEEE Transactions on Magnetics,* **MAG-19**, pp. 2635–2638, 1983.

Various shortcomings of the early version of PE2D were overcome by reprogramming and additions. True Newton iteration is now used, permanent magnet materials can be handled, and external currents are permitted in eddy current problems. (Commercial software)

Brauer, J. R.: Finite element calculation of eddy currents and skin effects. *IEEE Transactions on Magnetics,* **MAG-18**, pp. 504–509, 1982.

The AOS-Magnetic analysis system is capable of performing steady-state sinusoidal analysis of magnetically linear problems. To illustrate, the eddy currents in and around steel members of an arc furnace are computed. (Commercial software)

Coulomb, J. L.: A simple interactive finite element processor for teaching purposes. *Journal of Applied Physics,* **53**, pp. 8417–8419, 1982.

A small interactive CAD system is described, limited in capabilities but adequate for teaching purposes. The system is a restricted subset of the FLUX system, to which it bears the expected relationship. (Proprietary software)

Csendes, Z. J., Freeman, E. M., Lowther, D. A., Silvester, P. P.: Interactive computer graphics in magnetic field analysis and electric machine design. *IEEE Transactions on Power Apparatus and Systems,* **PAS-100**, pp. 2862–2869, 1981.

MagNet Eleven, an early version of the MagNet integrated program package, is described. First-order triangular elements are used, with preconditioned conjugate gradient solution, in a strongly interactive environment (PDP-11 minicomputer, refresh color graphics, and input tablet). Data input is directly in terms of a finite element mesh, through an interactive mesh editor. An interactive postprocessor is used, capable of vector arithmetic and analysis, and various forms of display. (Commercial software)

de Beer, A., Polak, S. J., Wachters, A. J. H., van Welij, J. S.: The use of PADDY for the solution of 3-D magnetostatic problems. *IEEE Transactions on Magnetics,* **MAG-18**, pp. 617–619, 1982.

PADDY is a three-dimensional field solution system similar in concept to MAGGY, but incorporating a number of additional features not dependent on three dimensions. The main features and characteristics of this program are described and illustrated by examples. (Proprietary software)

Diserens, N. J.: A space charge beam option for the PE2D and TOSCA packages. *IEEE Transactions on Magnetics,* **MAG-18**, pp. 362–366, 1982.

Space-charge forces can be optionally included in electric field analyses performed by PE2D, in order to permit particle trajectory plotting. A similar possibility exists for TOSCA. (Commercial software)

Girgis, R. S., Yannucci, D. A., Templeton, J. B.: Performance parameters of power transformers using 3-D magnetic field calculations. *IEEE Transactions on Power Apparatus and Systems,* **PAS-103**, pp. 2708–2713, 1984.

A good example of a classical process-oriented software system, in which the sequence of data presentation and computation is fixed, and postprocessing facilities are provided as a choice between predefined options. The system is specifically oriented to transformer design, largely through the types of postprocessing options available. (Proprietary software)

Hoburg, J. F., Davis, J. L.: A student-oriented finite element program for electrostatic potential problems. *IEEE Transactions on Education,* **E-26**, pp. 138–142, 1983.

A conventionally structured but rather limited set of programs useful for undergraduate teaching of field theory. The programs are essentially batch oriented and have limited interactivity, but graphic output is provided. (Proprietary software available for educational use)

Konrad, A., Silvester, P.: Scalar finite-element program package for two dimensional field problems. *IEEE Transactions on Microwave Theory and Techniques,* **MTT-19**, pp. 952–954, 1971.

Description of a publicly available program using triangular elements up to sixth order to solve Laplace's, Poisson's, or Helmholtz's equation, the latter for the region eigenvalues and eigenfunctions. (Publicly available software)

Konrad, A.: A linear accelerator cavity code based on the finite element method. *Computer Physics Communications,* **13**, pp. 349–362, 1978.

Description of a program filed with the CPC program library, based on the author's previous article (IEEE NS-20, 802, 1973; see *Postprocessing Techniques* above). (Publicly available software)

Lari, R. J., Turner, L. R.: Survey of eddy current programs. *IEEE Transactions on Magnetics,* **MAG-19**, pp. 2474–21477, 1983.

A survey of eddy current programs available commercially, distributed through private channels, or in the public domain. There are some major omissions (e.g., all of the CPC library), particularly from the latter, but fairly thorough descriptions are given of most programs.

Lowther, D. A.: A microprocessor based electromagnetic field analysis system. *IEEE Transactions on Magnetics,* **MAG-18**, pp. 351–356, 1982.

An early version of the MagNet system is described, designed to run on an LSI-11 microprocessor with suitable interactive graphics support. Examples of mesh generation and postprocessing are shown. (Commercial software)

MacNeal, R. H., McCormick, C. W.: The NASTRAN computer program for structural analysis. *Computers and Structures,* **1**, pp. 389–412, 1971.

A general description of NASTRAN, with emphasis on the types of algorithms and kinds of elements available. Some comments on the NASTRAN executive are also included. (Commercial civil engineering software)

Pillsbury, R. D., Jr.: NMLMAP—a two dimensional finite element program for transient or static linear or nonlinear magnetic field problems. *IEEE Transactions on Magnetics,* **MAG-18**, pp. 406–410, 1982.

An essentially batch-processing-oriented finite element program is described, of extensive capabilities and relatively classical software structure. (Proprietary software)

Pissanetzky, S.: Solution of three-dimensional, anisotropic, nonlinear problems of magnetostatics using two scalar potentials, finite and infinite multipolar elements and automatic mesh generation. *IEEE Transactions on Magnetics,* **MAG-18**, pp. 346–350, 1982.

Two scalar potentials, ordinary and reduced, are used in a finite element program to maintain numerical stability in anisotropic three-dimensional solutions. The solver program is flanked by a batch-oriented mesh construction package. (Part commercial software, part public domain)

Polak, S. J., de Beer, A., Wachters, A., van Welij, J. S.: MAGGY2, a program package for two-dimensional magnetostatic problems. *International Journal on Numerical Methods in Engineering,* **15**, pp. 113–128, 1980.

MAGGY2 (the in-house forerunner of MAGGY) is a large-scale package written in transportable Fortran and installed on various computers. It operates in a batch mode with Calcomp graphic output. The data input language requires only keyboard input; meshing details as well as geometric shapes are under user control. Topologically regular meshes are used. Both Gauss and conjugate gradient methods are available for equation solving in the Newton process. Various computational output options are available as well as graphic plots. (Commercial software)

Sabonnadière, J. C., Meunier, G., Morel, B.: FLUX: a general interactive finite elements package for 2D electromagnetic fields. *IEEE Transactions on Magnetics,* **MAG-18**, pp. 624–626, 1982.

The FLUX field solver as implemented under the Multics operating system is described very briefly. (Commercial software)

Simkin, J., Trowbridge, C. W.: Electromagnetics CAD using a single user machine (SUM). *IEEE Transactions on Magnetics,* **MAG-19**, pp. 2655–2658, 1983.

An interactive programming system for CAD is described. The system is housed on a sophisticated microcomputer workstation with high-resolution graphics and contains sophisticated input arrangements as well as a modest postprocessing subsystem. (Proprietary software)

Spence, R., Apperley, M.: The interactive-graphic man-computer dialogue in computer-aided circuit design. *IEEE Transactions on Circuits and Systems,* **CAS-24**, pp. 49–61, 1977.

Description of the major communication aspects of the Minnie circuit CAD system, including some observations on psychological aspects. (Proprietary circuit analysis software)

Tual, J. P.: 2-D computation problems in magnetic recording and printing. *IEEE Transactions on Magnetics,* **MAG-19**, pp. 2486–2490, 1983.

A finite element program is described, capable of performing magnetostatic calculations. It appears to be functionally comparable to commercially available analysis systems. The physical approximations admissible for recording analysis are stated. (Proprietary software)

Index